The Open University

Mathematics Foundation Course Unit 23

LINEAR ALGEBRA II

Prepared by the Mathematics Foundation Course Team

Correspondence Text 23

The Open University Press

Open University courses provide a method of study for independent learners through an integrated teaching system including textual material, radio and television programmes and short residential courses. This text is one of a series that make up the correspondence element of the Mathematics Foundation Course.

The Open University's courses represent a new system of university level education. Much of the teaching material is still in a developmental stage. Courses and course materials are, therefore, kept continually under revision. It is intended to issue regular up-dating notes as and when the need arises, and new editions will be brought out when necessary.

Further information on Open University courses may be obtained from The Admissions Office, The Open University, P.O. Box 48, Bletchley, Buckinghamshire.

The Open University Press
Walton Hall, Bletchley, Bucks

First Published 1971

Printed in Great Britain by
J W Arrowsmith Ltd, Bristol 3

SBN 335 01022 9

Contents

Objectives

The principal objective of this unit is to establish some general important results which will form the basis of applications to be considered in later units.

After working through this unit, you should be able to:

(i) explain the meanings of the terms:
   dimension of a vector space,
   morphism from one vector space to another,
   kernel of a morphism,
   matrix,
   identity matrix;

(ii) understand the proofs of the theorems about morphisms of vector spaces which are listed in section 23.1.6;

(iii) add and multiply (suitable) matrices;

(iv) form a scalar multiple of a matrix.

Note

Before working through this correspondence text, make sure you have read the general introduction to the mathematics course in the Study Guide, as this explains the philosophy underlying the whole course. You should also be familiar with the section which explains how a text is constructed and the meanings attached to the stars and other symbols in the margin, as this will help you to find your way through the text.

Structural Diagram

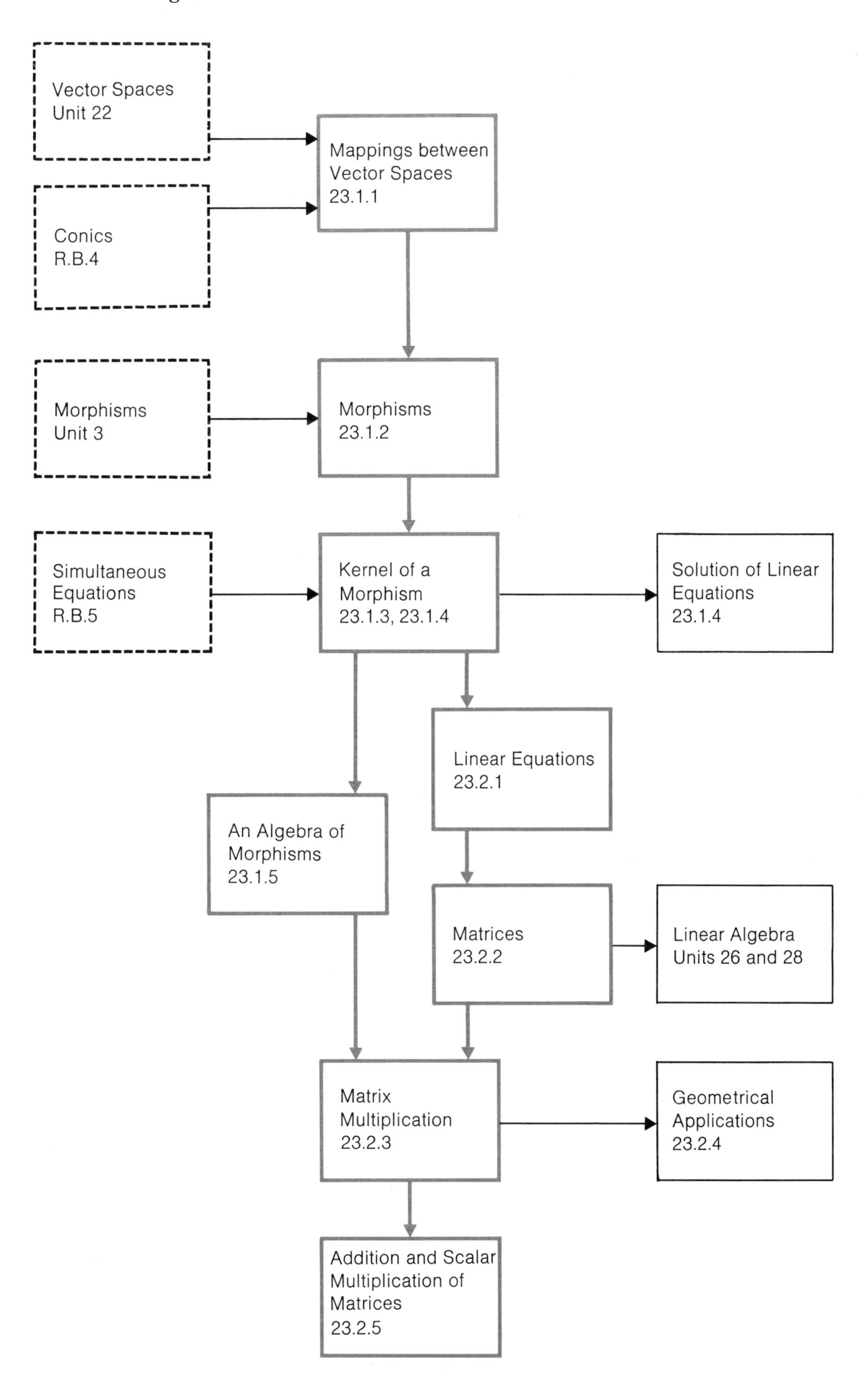

Glossary

Terms which are defined in this glossary are printed in CAPITALS.

Term	Definition	Page
EQUALITY OF MATRICES	Two MATRICES are EQUAL if they have the same numbers of rows and columns, and the corresponding elements are the same.	31
IDENTITY MATRIX	The IDENTITY MATRIX is a MATRIX with the same number of rows as columns, in which all the elements on the diagonal from the top left-hand corner to the bottom right-hand corner are 1's, and all other elements are 0's.	42
INVARIANT ELEMENT	An INVARIANT ELEMENT under a function f is an element a of the domain of f such that $f(a) = a$.	4
INVARIANT SET	An INVARIANT SET under a function f is a subset A of the domain of f such that $f(A) = A$.	4
KERNEL	The KERNEL of a MORPHISM is the set of elements in the domain which map to the zero element in the codomain of the morphism.	17
LINEAR EQUATION	A LINEAR EQUATION is an equation of the form $$a_1x_1 + a_2x_2 + \cdots + a_nx_n = 0,$$ where the a's are constants.	28
LINEAR TRANSFORMATION	A LINEAR TRANSFORMATION is a MORPHISM from one vector space to another.	9
MATRIX ($m \times n$)	A MATRIX ($m \times n$) is a rectangular array of numbers, having m rows and n columns.	30
MATRIX ADDITION	See page 43.	43
MATRIX MULTIPLICATION	See page 37.	37
MATRIX SUBTRACTION	See page 43.	43
MORPHISM	If L is a mapping from a vector space V to a vector space U such that $$L(\alpha_1\underset{\sim}{v}_1 + \alpha_2\underset{\sim}{v}_2) = \alpha_1L(\underset{\sim}{v}_1) + \alpha_2L(\underset{\sim}{v}_2),$$ for any $\alpha_1, \alpha_2 \in R$, $\underset{\sim}{v}_1, \underset{\sim}{v}_2 \in V$, then L is a MORPHISM from V to U.	9
MULTIPLICATION OF A MATRIX BY A SCALAR	See page 43.	43
NULL SPACE	The NULL SPACE of a MORPHISM is the KERNEL of the morphism.	17
SUBSPACE	If S is a subset of a vector space U, and S is itself a vector space, then S is a SUBSPACE of U.	13

Notation

		Page
$\underset{\sim}{i}, \underset{\sim}{j}, \underset{\sim}{k}$	Geometric vectors of length one in the directions of the x, y and z Cartesian axes respectively.	1
R^n	The set of all n-tuples of real numbers, i.e. the Cartesian product, $\underbrace{R \times R \times \cdots \times R}_{n \text{ terms}}.$	2
$\underset{\sim}{f}$	The function f, considered as an element of a vector space.	6
$+_v$	A temporary symbol for the operation of addition on a vector space V.	9
$\underset{\sim}{v}_0$	The zero element of the vector space V.	10
$\underset{\sim}{e}_1, \underset{\sim}{e}_2, \ldots, \underset{\sim}{e}_n$	$(1, 0, \ldots, 0)$, $(0, 1, \ldots, 0)$, $\ldots$, $(0, 0, \ldots, 1)$: These n-tuples form a basis for R^n.	29
$\begin{pmatrix} a_1 & a_2 & a_3 \\ b_1 & b_2 & b_3 \end{pmatrix}$	A matrix with two rows and three columns.	31
$\square$	A temporary symbol for the operation of multiplication of two matrices A and B (in that order) where B has just one column.	32
$*$	A temporary symbol for the operation of matrix multiplication.	35
R_θ	The matrix corresponding to the mapping which rotates the plane about the origin through an angle θ anti-clockwise, i.e. $\begin{pmatrix} \cos\theta & -\sin\theta \\ \sin\theta & \cos\theta \end{pmatrix}$	39
I	The identity matrix.	42
$\triangle$	A temporary symbol for the operation of matrix addition.	43

Bibliography

H. S. Wilf, *Calculus and Linear Algebra* (Harcourt, Brace and World 1966).
This book is easy to read and, as the title suggests, may be a help in the Calculus as well as Linear Algebra. The material relevant to this unit is covered in Chapter 6; we shall cover the first part of this chapter later in the course, so section 6.1 may be omitted at this stage.

F. Loonstra, *Introduction to Algebra* (McGraw-Hill 1967).
Written in a different style to the first book, this text is a more formal introduction to Algebra. It is very clearly written. The relevant chapter is again Chapter 6.

S. Lang, *Introduction to Linear Algebra* (Addison-Wesley 1970).
The approach is similar to ours in that vector spaces are introduced by way of geometrical vectors. The order in which matrices and morphisms are discussed is slightly different from ours, but nevertheless Chapters 4 and 5 cover the material of this unit.

23.1 MAPPINGS OF VECTOR SPACES

23.1

23.1.0 Introduction

23.1.0

Introduction

* *

In *Unit 22, Linear Algebra I* we considered the *set of geometric vectors*, and defined the terms *linear dependence* and *independence*, *dimension* and *basis*. We then extended our use of these terms from a geometric context to an algebraic one: we defined a *vector space*. The examples of vector spaces given in *Unit 22* are interesting because, although they look very different, they all belong under the same umbrella. However, just looking at examples does not give us any extra equipment for solving problems. In this unit we shall establish some general results for vector spaces.

You will recall that in *Unit 1* we began by mapping sets one to another, and we introduced the concept of a *function*. With this concept we found that we could think about problems which are more sophisticated and have wider application than simple arithmetic calculations. It is the same with vector spaces; the topic becomes richer and more interesting when we introduce mappings from one vector space to another.

23.1.1 Mapping One Vector Space to Another

23.1.1

Some mappings of vector spaces to vector spaces are simply equivalent to a change of notation.

Example 1

Example 1

We know that the set of geometric vectors forms a vector space, as does the set of all lists with three elements.

If $\underset{\sim}{v} = x\underset{\sim}{i} + y\underset{\sim}{j} + z\underset{\sim}{k}$, then the mapping

$$n : \underset{\sim}{v} \longmapsto \begin{pmatrix} x \\ y \\ z \end{pmatrix},$$

with domain the set of geometric vectors, simply gives us an alternative notation.

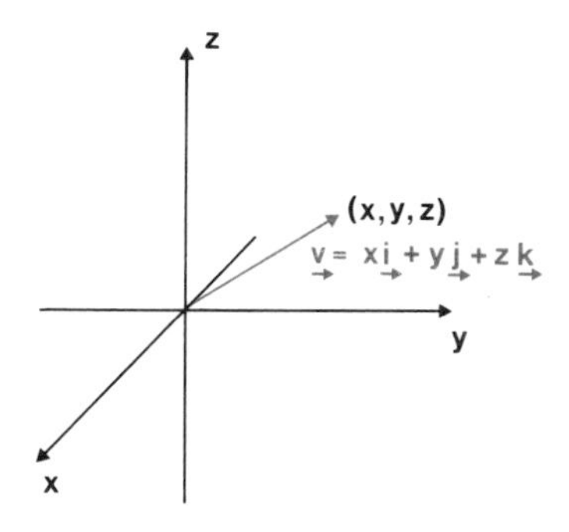

($\underset{\sim}{i}$, $\underset{\sim}{j}$ and $\underset{\sim}{k}$ are the unit geometric vectors in the directions of the x, y and z Cartesian axes respectively: see *Unit 22, Linear Algebra I*.)

Under this mapping we have

$$n: \underset{\sim}{i} \longmapsto \begin{pmatrix} 1 \\ 0 \\ 0 \end{pmatrix}$$

$$n: \underset{\sim}{j} \longmapsto \begin{pmatrix} 0 \\ 1 \\ 0 \end{pmatrix}$$

$$n: \underset{\sim}{k} \longmapsto \begin{pmatrix} 0 \\ 0 \\ 1 \end{pmatrix}$$

We can calculate $n(\underset{\sim}{v})$ for any geometric vector $\underset{\sim}{v}$ once we know the images of the base vectors. ■

Notice that if we want a mapping to define a new notation for the elements in its domain then the mapping must be one-one.

Another "notational" mapping is the mapping of ordered lists to ordered pairs defined by

$$\begin{pmatrix} x \\ y \end{pmatrix} \longmapsto (x, y),$$

or ordered lists to ordered triples defined by

$$\begin{pmatrix} x \\ y \\ z \end{pmatrix} \longmapsto (x, y, z),$$

which simply states the obvious fact that, given a co-ordinate system, we can represent a two-element list by a point in a plane and a three-element list by a point in three-dimensional space. We already have a name for the set of all ordered pairs of real numbers; we call it $R \times R$. This is usually shortened to R^2. R^3 is used to denote the set of all ordered triples of real numbers, and R^n is used to denote the set of all ordered n-tuples of real numbers. We have seen that R^n, with addition and multiplication by a scalar defined in the usual way, is a vector space of dimension n. We shall refer to this vector space as R^n, leaving the operations to be understood.

Example 2

Example 2

Consider the mapping of R^2 to R^2 defined by

$$f: (x, y) \longmapsto (2x, 2y).$$

What effect does this mapping have? One way to start to answer this question is to consider what happens to a few particular elements.

Take the element $(1, 1)$; then $f(1, 1) = (2, 2)$.

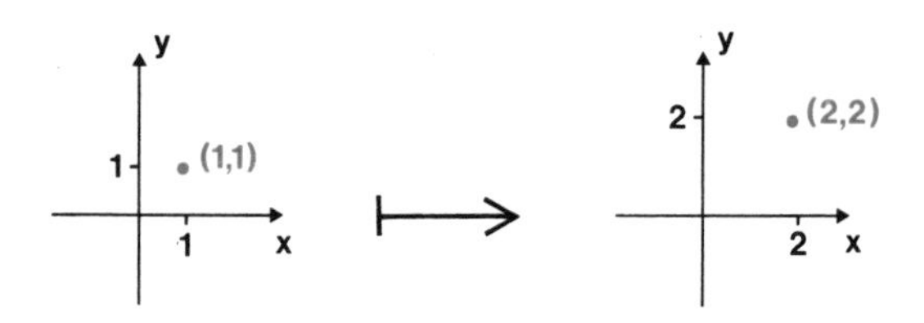

For the element (0, 1), we have $f(0, 1) = (0, 2)$.

In general, we see that the point P is mapped on to the point P' in the same direction away from the origin O, where $OP' = 2OP$. (Does any point remain unchanged?)

Consider a set of elements in the plane, such as

$$\{(x, y): x^2 + y^2 = 1\}.$$

These points lie on a circle of unit radius, and therefore their images under this mapping will lie on a circle of radius 2. We can verify this algebraically as follows. Suppose $(x, y) \longmapsto (u, v)$, then $u = 2x$ and $v = 2y$, and so if x and y satisfy the equation

$$x^2 + y^2 = 1,$$

u and v must satisfy the equation

$$\left(\frac{u}{2}\right)^2 + \left(\frac{v}{2}\right)^2 = 1$$

i.e. $$u^2 + v^2 = 4$$

Thus

$$\{(x, y): x^2 + y^2 = 1\} \longmapsto \{(u, v): u^2 + v^2 = 4\}$$
$$= \{(x, y): x^2 + y^2 = 4\}$$

(See RB4)

(We have re-written this set in terms of x and y so that we can draw the following diagram.)

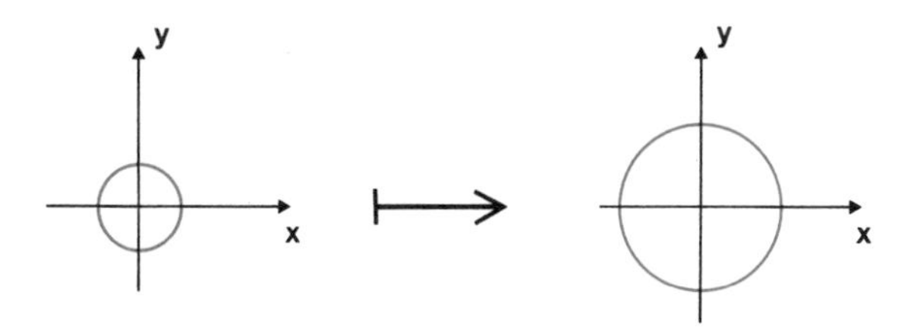

■

Example 3

Example 3

Consider the mapping of R^2 to R^2 defined by

$$(x, y) \longmapsto (-y, x).$$

Let us have a look at what happens to the base vectors (1, 0) and (0, 1). We have

$$(1, 0) \longmapsto (0, 1) \quad \text{and} \quad (0, 1) \longmapsto (-1, 0).$$

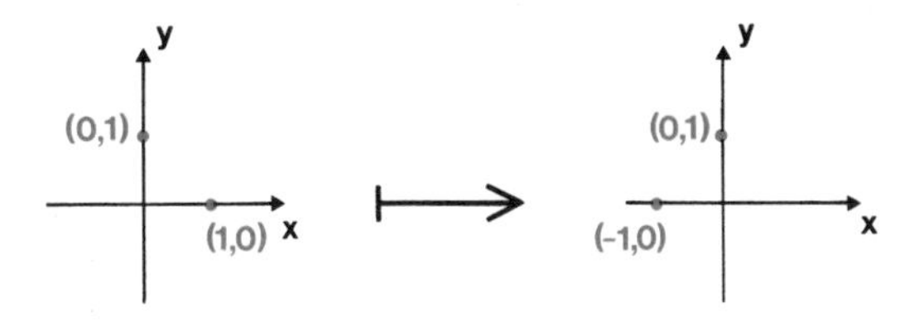

The mapping has the effect of rotating the base vectors through an angle $\frac{\pi}{2}$ counter-clockwise about the origin, and this is indeed the effect on the entire plane.

In this example, any circle centred at the origin maps onto itself. Every point of the set $\{(x, y): x^2 + y^2 = 1\}$ moves (for example, $(\frac{3}{5}, \frac{4}{5}) \longmapsto (-\frac{4}{5}, \frac{3}{5})$), but the set itself remains unchanged. ■

Example 3 provides a particular example of a simple but important mathematical concept. Given a function f of a set A to itself, then if, for instance,

$$f(a) = a \qquad (a \in A),$$

we say that a is an invariant element under f.

Definition 1
* * *

If the set S *as a whole* maps to itself, then S is called an invariant set under f, even if some or all of the elements of the subset S of A are moved under f. In the above example we have an invariant circle, i.e. an invariant set of points. The concept of invariance can be even more general. For example, under a translation of the plane, the distance between two points is invariant; that is, if P and Q map to P' and Q' respectively, then the length of PQ = length of $P'Q'$. Invariance is a very important and general concept in mathematics, and we shall have more to say about it in the remainder of this course and subsequently. In fact, we encountered invariance very early in the Foundation Course. When we are asked to solve an equation $f(x) = 0$ and we choose to do it iteratively, then we usually rearrange the equation in the form:

Definition 2
* * *

$$x = F(x);$$

that is, we look for the *invariant elements* under the function F. (See *Unit 2, Errors and Accuracy*.)

Exercise 1

Exercise 1
(1 minute)

Under the mapping

$$f: (x, y) \longmapsto (2x, 2y),$$

in Example 2, we have seen that circles centred at the origin are not invariant. But some lines are invariant. Which lines? ■

Example 4

Example 4

Consider the mapping of R^2 to R^2 defined by

$$(x, y) \longmapsto (-y, y).$$

From the previous two examples, it seems that, to see the effect of the mapping, it is probably worth seeing what happens to a set of base vectors. Again, to simplify the arithmetic, we choose the base vectors $(1, 0)$ and $(0, 1)$. We find that

$$(1, 0) \longmapsto (0, 0)$$

$$(0, 1) \longmapsto (-1, 1),$$

so that in terms of the corresponding geometric vectors $\underset{\sim}{i}$ and $\underset{\sim}{j}$, we have that $\underset{\sim}{j}$ maps to $-\underset{\sim}{i} + \underset{\sim}{j}$, but $\underset{\sim}{i}$ maps to the zero vector.

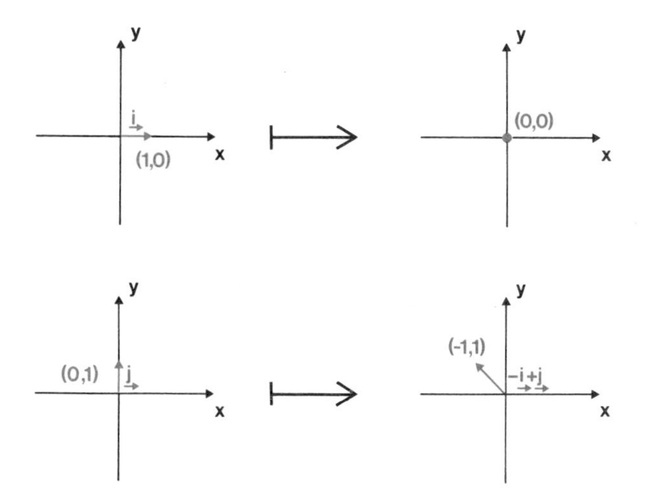

We shall look at this mapping in a little more detail. What happens to the x-axis? On this axis y is zero and so every point on the x-axis maps to $(0, 0)$: the entire x-axis shrinks into the origin. What about the line $y = 1$? Every point on this line maps to the point $(-1, 1)$.

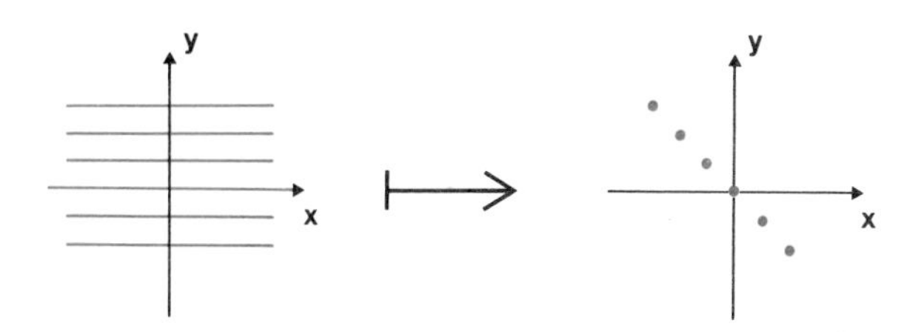

In fact, the entire plane maps on to the line whose equation is

$$x + y = 0$$

In terms of geometric vectors, every element in the image set can be represented in terms of the vector $-\underset{\sim}{i} + \underset{\sim}{j}$.

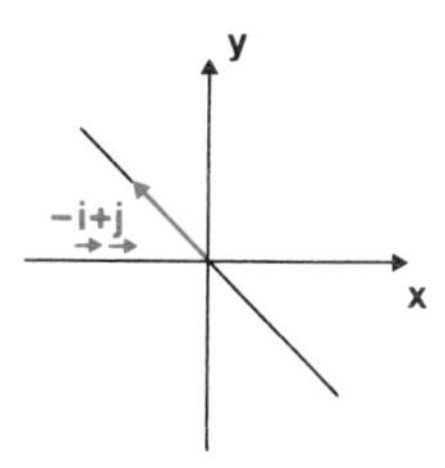

The image set has dimension 1, and so the effect of the mapping is to "lose" a dimension from our vector space. This is equivalent to saying that this mapping is many-one. If we start with a point, P say, in the plane and map it to a point Q on the line $x + y = 0$, then we cannot map back to the original point P. This is because the point Q on the line $x + y = 0$ corresponds to the *whole* of the line parallel to the x-axis through P; this line is the equivalence class containing the element P. (In *Unit 19, Relations* we showed that any many-one mapping defines an equivalence relation on its domain. This is an example. The relation ρ is $(a, b)\, \rho\, (c, d)$ if $b = d$.) ■

(continued on page 6)

Solution 1

Solution 1

Straight lines which pass through the origin. ■

(*continued from page 5*)

Discussion

We have chosen these geometric examples to give you a visualization of the sort of mappings we are going to consider. One of the pleasant features of linear algebra is that by considering a geometric situation we can often throw light on non-geometric situations (and vice versa). Thus, a non-geometric analogue of the last case, where we "lost" a dimension, is provided by the following example.

Example 5

Example 5

Let P_3 be the vector space of all polynomial functions of degree 3 or less. (This space has dimension 4, and a set of base vectors is given below.) The differentiation operator:

$$D: \underline{p} \longmapsto D\underline{p} \qquad (\underline{p} \in P_3)$$

maps the space P_3 to the space P_2 (which has dimension 3). In this case we could take as a basis for P_3 the set of functions:

$$\left.\begin{array}{ll} \underline{f}_0 : x \longmapsto 1 & (x \in R) \\ \underline{f}_1 : x \longmapsto x & (x \in R) \\ \underline{f}_2 : x \longmapsto x^2 & (x \in R) \\ \underline{f}_3 : x \longmapsto x^3 & (x \in R). \end{array}\right\} \text{Basis for } P_3$$

This set of functions maps to the set

$$\begin{array}{ll} \underline{f}'_0 : x \longmapsto 0 & (x \in R) \end{array}$$

$$\left.\begin{array}{ll} \underline{f}'_1 : x \longmapsto 1 & (x \in R) \\ \underline{f}'_2 : x \longmapsto 2x & (x \in R) \\ \underline{f}'_3 : x \longmapsto 3x^2 & (x \in R). \end{array}\right\} \text{Basis for } P_2$$

Like the previous example, this mapping is many-one. (For example, all the functions of the form:

$$\underline{f}: x \longmapsto x^2 + a \qquad (x \in R),$$

where $a \in R$, map to the function $\underline{f}'_2$.)

Note that $\underline{f}'_0$ is the *zero vector* in P_2. It cannot belong to any basis for P_2, since a basis must be a linearly independent set of three vectors. For consider the set $\{\underline{f}'_0, \underline{g}, \underline{h}\}$, where $\underline{g}, \underline{h} \in P_2$. Since

$$\alpha \underline{f}'_0 + 0\underline{g} + 0\underline{h} = \underline{f}'_0,$$

where α is any *non-zero* real number, we see that any set of three elements containing $\underline{f}'_0$ is linearly *dependent*. ■

Exercise 2

Exercise 2
(2 minutes)

We seem to be pinning a lot of faith in choosing a convenient basis. Is the choice of basis important? We shall resolve this difficulty later, but one point can be considered here.

In Example 5, instead of $\underline{f}_0$ and $\underline{f}_1$, we could choose $\underline{g}_0$ and $\underline{g}_1$, where

$$\underline{g}_0 : x \longmapsto 1 + x \qquad (x \in R)$$

$$\underline{g}_1 : x \longmapsto 1 - x \qquad (x \in R),$$

and then none of the base vectors maps to the zero vector under D. Is $\{\underline{g}'_0, \underline{g}'_1, \underline{f}'_3\}$ a basis for P_2? ■

Exercise 3

Exercise 3
(2 minutes)

Fill in the gaps in the solution to the following problem.

Problem

If T is a mapping from R^2 to R^2, and

$$T:(x, y) \longmapsto \left(\frac{x}{2}, \frac{y}{3}\right),$$

describe the image of the ellipse

$$\{(x, y): 9x^2 + 4y^2 = 36\}.$$

Solution of Problem

Put $u = \frac{x}{2}$ and $v = \frac{y}{3}$

Since x and y satisfy the equation

$$9x^2 + 4y^2 = 36$$

u and v satisfy the equation

$$\boxed{\qquad\qquad = 36} \qquad \text{(i)}$$

i.e.

$$\boxed{\qquad\qquad = 1} \qquad \text{(ii)}$$

i.e.

$$\{(x, y): 9x^2 + 4y^2 = 36\} \longmapsto \boxed{\{(u, v): \qquad \}} \qquad \text{(iii)}$$

The image set corresponds to a circle centred at

$\boxed{\qquad}$ and with radius $\boxed{\qquad}$ (iv)

■

Exercise 4

Exercise 4
(4 minutes)

(i) By choosing a suitable basis and finding its image, or otherwise, describe the effect of the following mappings of R^2 to R^2.

(a) $T_1:(x, y) \longmapsto (y, x)$

(b) $T_2:(x, y) \longmapsto \left(\frac{x}{\sqrt{2}} - \frac{y}{\sqrt{2}}, \frac{x}{\sqrt{2}} + \frac{y}{\sqrt{2}}\right)$

(ii) Find an element in R^2 which is unchanged under the mapping T_1. ■

Solution 2

Solution 2

$$g'_0 : x \longmapsto 1 \qquad (x \in R)$$

$$g'_1 : x \longmapsto -1 \qquad (x \in R).$$

Although none of the base vectors is mapped to the zero vector, $\{g'_0, g'_1\ f'_3\}$ is linearly *dependent*, since

$$1g'_0 + (-1)g'_1 + 0f'_3 = f'_0;$$

g'_0 and g'_1 cannot both belong to the *same* basis because one is a scalar multiple of the other. ■

Solution 3

Solution 3

(i) $9(2u)^2 + 4(3v)^2 = 36$
(ii) $u^2 + v^2 = 1$
(iii) $\{(u, v) : u^2 + v^2 = 1\}$
(iv) the origin; 1

Thus the effect of the mapping is to transform the ellipse to the circle as shown below.

(See RB4)

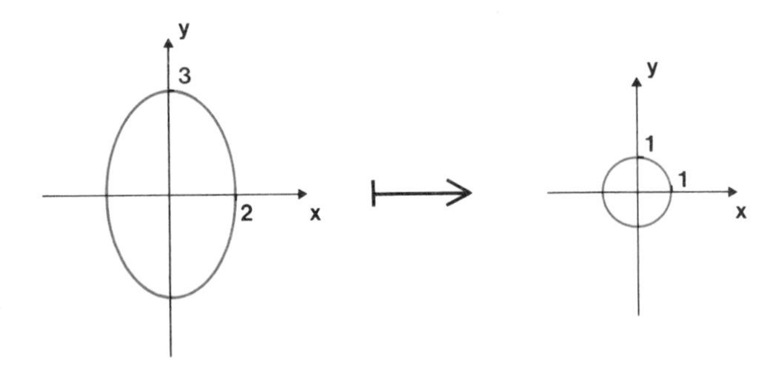

■

Solution 4

Solution 4

(i) If we choose the basis $\{(1, 0), (0, 1)\}$, we have:
(a) $T_1 : (1, 0) \longmapsto (0, 1)$
$T_1 : (0, 1) \longmapsto (1, 0)$

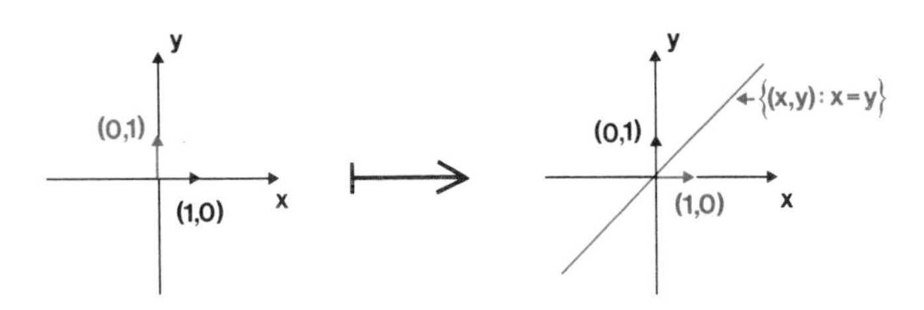

The effect of T_1 is to reflect the points of the plane in the line with equation $y = x$ (we are simply interchanging the x and y co-ordinates).

(b) $T_2 : (1, 0) \longmapsto \left(\frac{1}{\sqrt{2}}, \frac{1}{\sqrt{2}}\right)$

$T_2 : (0, 1) \longmapsto \left(-\frac{1}{\sqrt{2}}, \frac{1}{\sqrt{2}}\right)$

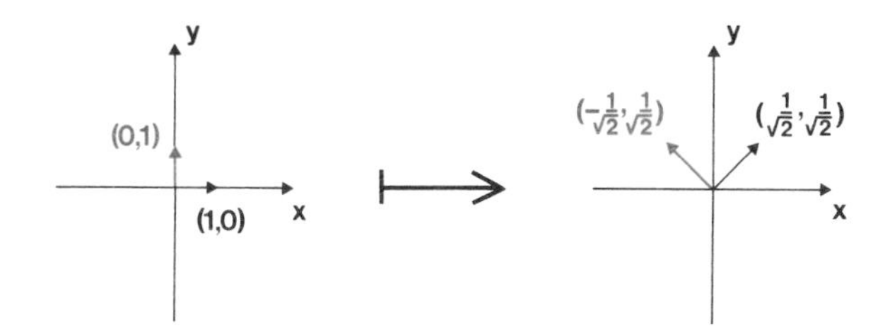

The effect of T_2 is to rotate the plane about the origin through an angle $\frac{\pi}{4}$. (If you are not convinced, find the images of some more points.)

(ii) Any scalar multiple of (1, 1) is an invariant element under T_1. In terms of geometric vectors, any vector in the direction of the line with equation $x = y$ is invariant. All such geometric vectors are scalar multiples of $\underset{\sim}{i} + \underset{\sim}{j}$. ■

23.1.2 Morphisms

23.1.2

Discussion
* *

There are many interesting and useful results concerning mappings of vector spaces, but the most fruitful field of study consists of those mappings which are *homomorphisms* or *isomorphisms* (and therefore, of necessity, *functions*).

If V is a vector space, the two operations we have defined on the elements of V are addition of vectors and multiplication of a vector by a scalar. Suppose a function T maps a vector space V, with an addition operation $+_v$, to a vector space U, with an addition operation $+_u$. The additive structure will be preserved if, for any vectors $\underset{\sim}{v}_1$ and $\underset{\sim}{v}_2$ in V,

$$T(\underset{\sim}{v}_1 +_v \underset{\sim}{v}_2) = T(\underset{\sim}{v}_1) +_u T(\underset{\sim}{v}_2)$$

Since we have been abusing the symbol + throughout this unit (we have defined all sorts of methods of addition and called them all +), we shall continue to do so, and we drop the suffices u and v from the addition symbols. We then have

$$T(\underset{\sim}{v}_1 + \underset{\sim}{v}_2) = T(\underset{\sim}{v}_1) + T(\underset{\sim}{v}_2)$$

Equation (1)

as the condition that T should be a morphism for the addition operations.

For the other operation we require that

$$T(\alpha\underset{\sim}{v}) = \alpha T(\underset{\sim}{v}),$$

Equation (2)

for any real number α and any vector $\underset{\sim}{v} \in V$.

Equations (1) and (2) are the conditions that T should be a morphism from the vector space V to the vector space U.*

The two equations can be combined to give the following equation:

$$T(\alpha_1\underset{\sim}{v}_1 + \alpha_2\underset{\sim}{v}_2) = \alpha_1 T(\underset{\sim}{v}_1) + \alpha_2 T(\underset{\sim}{v}_2),$$

Equation (3)

for any real numbers α_1, α_2, and any vectors $\underset{\sim}{v}_1$ and $\underset{\sim}{v}_2 \in V$.

Definition 1
*

In many books, a mapping of a vector space to a vector space is called a *transformation*, and when the mapping is a morphism it is called a linear transformation. This is another example of calling a particular type of mapping by a special name. (We have already used the word *operator*.)

* Previously (see *Unit 3, Operations and Morphisms*, page 25), we said that a morphism f was a mapping from A to $f(A)$. Here, $T(V)$ may be a proper subset of U. (In fact, if T is a morphism, $T(V)$ *is* a vector space, but we have not proved this yet.)

Exercise 1

Exercise 1
(3 minutes)

Which of the following mappings are morphisms? (Take the operations in the various vector spaces to be the usual ones.)

(i) The mapping of R^2 to R^2 such that

$$T:(x_1, x_2) \longmapsto (x_2, x_1)$$

(ii) The mapping of R^2 to R^2 such that

$$T:(x_1, x_2) \longmapsto (x_1^2, x_2^2)$$

(iii) The mapping of the set of all geometric vectors in a plane to R such that

$$T:\underset{\to}{x} \longmapsto \underset{\to}{a} \cdot \underset{\to}{x},$$

where $\underset{\to}{a}$ is a given geometric vector, and the dot stands for the inner product as defined in *Unit 22*.

(iv) The mapping of the set of all polynomial functions of degree n or less to itself such that

$$T:\underset{\sim}{p} \longmapsto \text{the derived function of } \underset{\sim}{p}.$$

(v) The mapping of the set of all real functions, which are twice-differentiable at all points in R, to the set of all real functions, such that

$$T:\underset{\sim}{f} \longmapsto 2D^2\underset{\sim}{f} + D\underset{\sim}{f} + 3\underset{\sim}{f}$$ ■

Exercise 2

Exercise 2
(4 minutes)

Let L be a morphism from a vector space V to a vector space U.

Complete the gaps in the proof of the following theorem.

THEOREM

If the zero element of V is $\underset{\sim}{v}_0$ (i.e. $\underset{\sim}{v}_0$ is the element for which $\underset{\sim}{v} + \underset{\sim}{v}_0 = \underset{\sim}{v}$ for any $\underset{\sim}{v} \in V$), and if $\underset{\sim}{u}_0$ is the zero element of U, then $L(\underset{\sim}{v}_0) = \underset{\sim}{u}_0$.

PROOF

Since $\underset{\sim}{v} + \underset{\sim}{v}_0 = \underset{\sim}{v}$

$$L(\underset{\sim}{v} + \underset{\sim}{v}_0) = L(\boxed{}) \qquad \text{(a)}$$

But L is a morphism, so

$$L(\underset{\sim}{v} + \underset{\sim}{v}_0) = L(\boxed{}) + L(\boxed{}) \qquad \text{(b)}$$

From (i) and (ii), $L(\underset{\sim}{v}) + L(\underset{\sim}{v}_0) = L(\underset{\sim}{v})$,

so subtracting $L(\underset{\sim}{v})$ from both sides, we see that

$L(\underset{\sim}{v}_0)$ is the $\boxed{}$ element of U. (c)

Confirm this result for each of the mappings which are morphisms in Exercise 1. ■

Main Text
* * *

Exercise 2 shows that under a morphism the zero element in the domain vector space is mapped to the zero element in the codomain vector space. We shall use this result in proving the following theorem, which is fundamental to the study of morphisms of vector spaces.

THEOREM

Theorem
* * *

If L is a morphism from a vector space V to a vector space U, then $L(V)$ is a subset of U which is *itself* a vector space.

($L(V)$ may be U or a proper subset of U.)

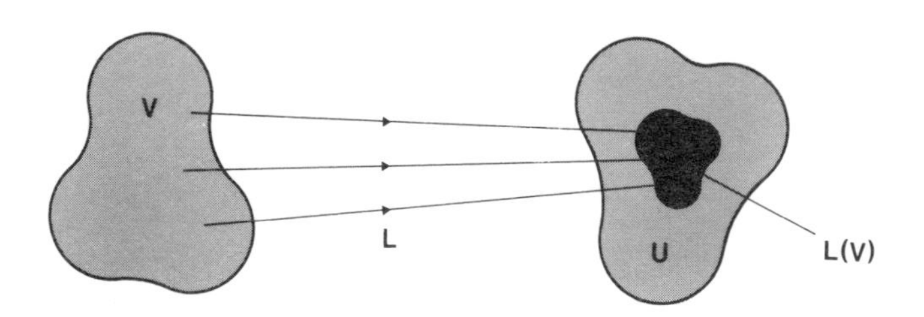

METHOD OF PROOF

Discussion
* *

We have to prove that the set $L(V)$, with the operations of the vector space U, satisfies the vector space axioms listed in *Unit 22*. To save you the trouble of referring back, we list the axioms for a vector space V; $\underset{\sim}{v}, \underset{\sim}{v}_1, \underset{\sim}{v}_2, \underset{\sim}{v}_3$ are any elements of V, and α and β are any real numbers:

1 $\underset{\sim}{v}_1 + \underset{\sim}{v}_2 \in V$ and is unique.
2 $\underset{\sim}{v}_1 + (\underset{\sim}{v}_2 + \underset{\sim}{v}_3) = (\underset{\sim}{v}_1 + \underset{\sim}{v}_2) + \underset{\sim}{v}_3$
3 $\underset{\sim}{v}_1 + \underset{\sim}{v}_2 = \underset{\sim}{v}_2 + \underset{\sim}{v}_1$
4 There is an element, $\underset{\sim}{v}_0$, in V, such that

$$\underset{\sim}{v} + \underset{\sim}{v}_0 = \underset{\sim}{v}$$

5 $\alpha\underset{\sim}{v} \in V$
6 $\underset{\sim}{v} + (-1)\underset{\sim}{v} = \underset{\sim}{v}_0$
7 $\alpha(\underset{\sim}{v}_1 + \underset{\sim}{v}_2) = (\alpha\underset{\sim}{v}_1) + (\alpha\underset{\sim}{v}_2)$
8 $(\alpha + \beta)\underset{\sim}{v} = \alpha\underset{\sim}{v} + \beta\underset{\sim}{v}$
9 $(\alpha\beta)\underset{\sim}{v} = \alpha(\beta\underset{\sim}{v})$
10 $1 \times \underset{\sim}{v} = \underset{\sim}{v}$

Axioms 2, 3, 6, 7, 8, 9 and 10 are statements about all elements of a vector space, and since U is a vector space we do not have to check these axioms for $L(V)$. On the other hand, axiom 4 is a statement that a particular kind of element (the zero element) belongs to a vector space. Clearly, the zero element of U will not belong to *every* subset of U, so we have to prove that it belongs to the particular subset $L(V)$. Axioms 1 and 5 concern *closure*. If $L(V)$ is to be a vector space, then any combination of elements in $L(V)$ must give resulting elements still in $L(V)$. This again is not necessarily true for *any* subset of U, so we must check it for $L(V)$.

PROOF

Main Text
* *

We have to prove that axioms 1, 4 and 5 hold for $L(V)$. We have three pieces of information:

(i) V is a vector space;
(ii) U is a vector space;
(iii) L is a morphism.

We have used (ii) to dispose of axioms 2, 3, 6, 7, 8, 9 and 10, but we have not yet used (i) and (iii).

We have proved that axiom 4 holds for $L(V)$: in Exercise 2 we proved that under a morphism the zero element $\underset{\sim}{v}_0$ in the domain V maps to the zero element $\underset{\sim}{u}_0$ in the codomain U. So the zero element of U belongs to $L(V)$.

Let us have a look at axiom 1; we must show that $L(V)$ is closed under addition. If $\underset{\sim}{u}_1$ and $\underset{\sim}{u}_2$ are any elements of $L(V)$, then there are elements

(continued on page 13)

Solution 1

Solution 1

(i)
$$T((x_1, x_2) + (y_1, y_2)) = T(x_1 + y_1, x_2 + y_2)$$
$$= (x_2 + y_2, x_1 + y_1)$$
$$= (x_2, x_1) + (y_2, y_1)$$
$$= T(x_1, x_2) + T(y_1, y_2),$$

and

$$T(\alpha(x_1, x_2)) = T(\alpha x_1, \alpha x_2)$$
$$= (\alpha x_2, \alpha x_1)$$
$$= \alpha(x_2, x_1)$$
$$= \alpha T(x_1, x_2),$$

so T is a morphism.

(ii)
$$\alpha T(x_1, x_2) = \alpha(x_1^2, x_2^2) = (\alpha x_1^2, \alpha x_2^2),$$

and

$$T(\alpha(x_1, x_2)) = T(\alpha x_1, \alpha x_2) = (\alpha^2 x_1^2, \alpha^2 x_2^2)$$

Equation (2) is not satisfied, so T is not a morphism.

(iii)
$$T(\underset{\sim}{x} + \underset{\sim}{y}) = \underset{\sim}{a} \cdot (\underset{\sim}{x} + \underset{\sim}{y})$$
$$= \underset{\sim}{a} \cdot \underset{\sim}{x} + \underset{\sim}{a} \cdot \underset{\sim}{y} \ (\cdot \text{ is distributive over } +)$$
$$= T(\underset{\sim}{x}) + T(\underset{\sim}{y})$$

and

$$T(\alpha \underset{\sim}{x}) = \underset{\sim}{a} \cdot \alpha \underset{\sim}{x}$$
$$= \alpha(\underset{\sim}{a} \cdot \underset{\sim}{x})$$
$$= \alpha T(\underset{\sim}{x})$$

so T is a morphism.

(iv) T is a morphism
(v) T is a morphism
} These results follow directly from the properties of D. ■

Solution 2

Solution 2

(a) $\underset{\sim}{v}$
(b) $\underset{\sim}{v}, \underset{\sim}{v}_0$
(c) zero

For the morphisms of Exercise 1, we have:

(i) $(0, 0) \longmapsto (0, 0)$
(iii) $\underset{\sim}{0} \longmapsto 0$
(iv) $(x \longmapsto 0) \longmapsto (x \longmapsto 0)$
(v) $(x \longmapsto 0) \longmapsto (x \longmapsto 0)$

You may have noticed that for many-one mappings, the zero vector is not the only vector which maps to the zero vector. For example, in (iv) we also have $(x \longmapsto k)$ mapping to $(x \longmapsto 0)$, where k is any real number. ■

(*continued from page 11*)

$\underset{\sim}{v}_1$ and $\underset{\sim}{v}_2$ in V such that

$$\underset{\sim}{u}_1 = L(\underset{\sim}{v}_1),$$
$$\underset{\sim}{u}_2 = L(\underset{\sim}{v}_2).$$

Then

$$\begin{aligned}\underset{\sim}{u}_1 + \underset{\sim}{u}_2 &= L(\underset{\sim}{v}_1) + L(\underset{\sim}{v}_2)\\ &= L(\underset{\sim}{v}_1 + \underset{\sim}{v}_2) \text{ (because } L \text{ is a morphism)}\\ &= L(\underset{\sim}{v}_3),\end{aligned}$$

where $\underset{\sim}{v}_3$ is an element of V (by axiom 1 for the vector space V); $L(\underset{\sim}{v}_3)$ is an element of $L(V)$, so $\underset{\sim}{u}_1 + \underset{\sim}{u}_2$ belongs to $L(V)$, and $L(V)$ is closed.

The other closure axiom, number 5, is easily checked.

If

$$\underset{\sim}{u} = L(\underset{\sim}{v}),$$

then

$$\begin{aligned}\alpha\underset{\sim}{u} &= \alpha L(\underset{\sim}{v})\\ &= L(\alpha\underset{\sim}{v}) \quad \text{(because } L \text{ is a morphism).}\end{aligned}$$

Therefore $\alpha\underset{\sim}{u} \in L(V)$.
This completes the proof.

If a subset of a vector space U is itself a vector space, then we call it a vector subspace of U.

Definition 2
* * *

Summary

Summary
*

So far in the text, we have considered mappings of one vector space to another, and we have concentrated our attention on those mappings which are morphisms. A morphism has the property that the image set itself is a vector space: we have seen that the properties of commutativity, associativity and the zero element are carried over from the domain to its image set by a morphism.

An interesting feature of *some* of the morphisms we have met is that they map a vector space on to an image set which has a lower dimension. For example, we have had mappings of planes to lines, polynomials of degree n or less to polynomials of degree $n - 1$ or less, and so on. Two questions arise. What has happened to the "lost" dimensions? Can we predict in advance when we are going to "lose" a dimension? We shall look at these questions in the next section.

Exercise 3

Exercise 3
(3 minutes)

(i) The mapping from R^2 to R^2 defined by

$$L:(x_1, x_2) \longmapsto (-x_2, x_2)$$

is a morphism. Prove directly by verifying the axioms that $L(R^2)$ is a vector space.

(ii) The mapping from R^2 to R^2 defined by

$$T:(x_1, x_2) \longmapsto (x_1^2, x_2)$$

is not a morphism. Show that $T(R^2)$ is not a vector space by finding an axiom which is not satisfied.

(Note that we have *not* proved that if $T: V \longmapsto U$ is *not* a morphism, then $T(V)$ is *not* a vector space. For a general mapping T, we do not know anything about $T(V)$ if T is not a morphism.) ■

Solution 3

Solution 3

(i) As in the proof of the theorem, we need to check axioms 1, 4 and 5 only.
Axiom 1 The elements of $L(R^2)$ are of the form

$$(-a, a), a \in R,$$

and

$$(-a, a) + (-b, b) = (-(a + b), a + b),$$

and so axiom 1 is satisfied.
Axiom 4 Consider the element (0, 0).

$$L((0, 0)) = (0, 0),$$

so that $(0, 0) \in L(R^2)$, and axiom 4 is satisfied.
Axiom 5 $(-a, a)$ is any element of $L(R^2)$

$$\alpha(-a, a) = (-\alpha a, \alpha a),$$

and so $\alpha(-a, a) \in L(R^2)$: axiom 5 is satisfied.

(ii) The only axioms which *may* not be satisfied are 1, 4 and 5. Of these only 5 is not satisfied.

$$\alpha(x_1^2, x_2) = (\alpha x_1^2, \alpha x_2)$$

and if α is negative, αx_1^2 is also negative, and so cannot be written as the square of a real number. ■

23.1.3 The Kernel

23.1.3

Discussion
* *

Let us have another look at Example 23.1.1.4 in which we saw that the morphism

$$L:(x, y) \longmapsto (-y, y) \qquad ((x, y) \in R^2)$$

maps the plane to a line. First, we looked at a particular basis, and saw that one of the base vectors mapped to the zero element (0, 0) in the codomain. We then investigated the mapping by looking to see what happened to particular subsets of the plane. We saw that any line parallel to the x-axis mapped to a single point.

This raises two questions. Firstly, is it significant that we lose *one* basis vector and we lose *one* dimension? Secondly, it is all very well in this simple case to pick out a few significant subsets that tell us such a lot. We picked them out because we knew their properties. Consider now the images of the lines parallel to the y-axis in the domain. *Any* such line maps to the *entire* image set, for suppose we take the line for which $x = a$, then

$$L:(a, y) \longmapsto (-y, y) \qquad (y \in R).$$

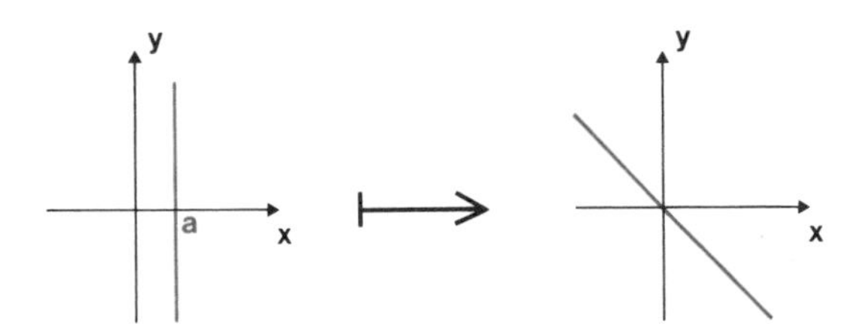

By considering certain subsets of the domain, we find that we can obtain information about L. Is there any particular subset which we can most profitably consider? That is, can we describe a subset in the domain which will give us information about L in a form which we can interpret easily? If so, can we extract any general feature which will help us with other examples?

The clue is in our observation about the loss of a basis vector. The vector $(1, 0)$ maps to $(0, 0)$, but of course it is not only this vector which "shrinks" to zero—every multiple of $(1, 0)$ maps to $(0, 0)$. (Because the mapping is a morphism, $L(\alpha(1, 0)) = \alpha L(1, 0) = (0, 0)$.) So a whole set maps to $(0, 0)$. Why consider this particular set? Try the next exercise.

Exercise 1

Exercise 1
(1 minute)

The vectors $(0, 1)$ and $(2, 2)$ form a basis for R^2. Calculate $L(0, 1)$ and $L(2, 2)$, where L is the mapping we have been discussing:

$$L:(x, y) \longmapsto (-y, y) \qquad ((x, y) \in R^2).$$

What happens to the linear independence of $(0, 1)$ and $(2, 2)$? ■

Discussion
* *

Exercise 1 shows us that, in our original basis, the choice of a vector which mapped to $(0, 0)$ was purely fortuitous. (See page 4.) It may so happen, as in this exercise, that none of the base vectors maps to $(0, 0)$, even though we "lose" a dimension. But here the whole set $\{(\alpha, 0): \alpha \in R\}$ maps to $(0, 0)$, whether or not one of its elements is in the basis.

It seems, then, that the set which maps to $(0, 0)$ tells us something about the "lost" dimension. In this case the set which maps to $(0, 0)$ has dimension 1 (every element of the set can be obtained as a scalar multiple of $(1, 0)$), and we lose just one dimension. Let us have a look at two more examples, one where we again lose one dimension and one where we lose more than one dimension. We shall again take geometrical examples because geometrical situations are easy to visualize.

Example 1

Example 1

The mapping

$$L:(x, y, z) \longmapsto (x, y, 0)$$

is a morphism of R^3 to R^3. The image of any point P is the point at the foot of the perpendicular from P to the plane with equation $z = 0$. Thus the domain maps to a plane.

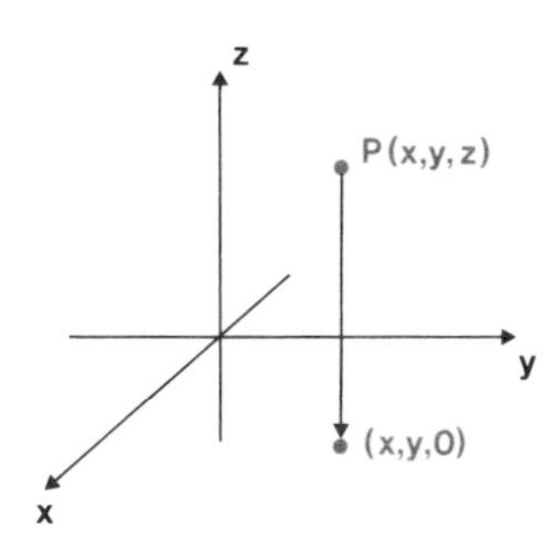

This morphism maps a 3-dimensional space to a 2-dimensional space: we lose one dimension. The set which maps to the zero element $(0, 0, 0)$ in the codomain is the set $\{(0, 0, z): z \in R\}$, that is, the z-axis. This set is itself a vector space and its dimension is one, the same number as the number of "lost" dimensions. ■

(*continued on page 16*)

Solution 1

Solution 1

$$L(0, 1) = (-1, 1) \neq (0, 0)$$

and

$$(2, 2) = (-2, 2) \neq (0, 0),$$

but

$$-2L(0, 1) + L(2, 2) = (0, 0)$$

Although neither vector maps to zero, the pair of linearly *independent* vectors maps to a pair of *dependent* vectors. So although the original vectors form a basis for R^2, their images do not. ■

(*continued from page 15*)

Example 2

Example 2

The mapping

$$L:(x, y, z) \longmapsto (x, 2x, 0)$$

is a morphism of R^3 to R^3. The image of the point (x, y, z) depends only on its x-co-ordinate. Thus (1, 2, 3), (1, 6, 7), (1, 6, 99) all map to the point (1, 2, 0). Every point in the plane with equation $x = 1$ maps to this point.

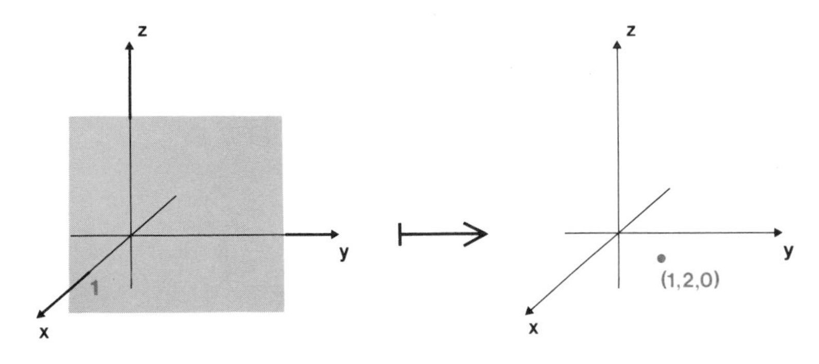

Similarly, every point on the plane with equation $x = 2$ maps to the point (2, 4, 0) – and so on. Every plane perpendicular to the x-axis maps to a point on the line defined by the equations:

$$2x - y = 0$$

$$z = 0,$$

and the entire three-dimensional space maps to this complete line.

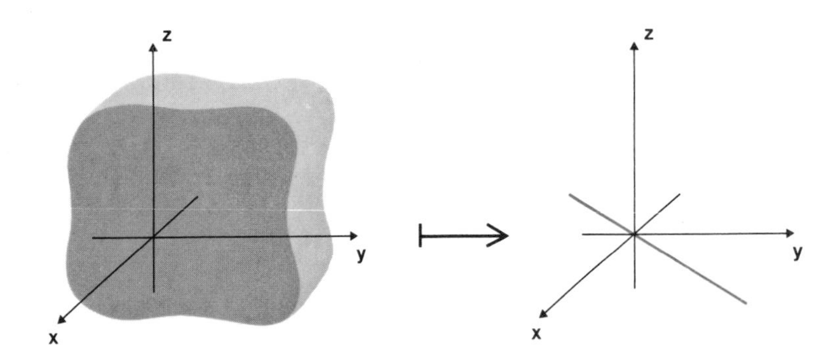

Thus

$$L(R^3) = \{(x, y, z): 2x - y = 0, z = 0\}$$

The three-dimensional domain maps to a space of dimension 1. We seem to have lost two dimensions.

Which set maps to the zero element? In this case the zero element is $(0, 0, 0)$, and the set which maps to $(0, 0, 0)$ is the set $\{(x, y, z) : x = 0\}$, i.e. the yz-plane; this is itself a vector space and has dimension two. Notice that in this simple case we can use the "basis argument" again — if we can find an appropriate basis. Taking the set of vectors $\{(1, 0, 0), (0, 1, 0), (0, 0, 1)\}$ as a basis, we see that both $(0, 1, 0)$ and $(0, 0, 1)$ map to $(0, 0, 0)$, and so we "lose" two base vectors. In fact we "lose" any vector which can be expressed as a linear combination of these two vectors (i.e. the points in the plane with equation $x = 0$), because

$$L(\alpha(0, 1, 0) + \beta(0, 0, 1))$$
$$= \alpha L(0, 1, 0) + \beta L(0, 0, 1) = (0, 0, 0),$$

since L is a morphism. ■

Main Text
* * *

We have seen that *the subset of the domain which maps to the zero element in the codomain* plays an important part, so we now give it a name.

If L is a morphism of a vector space V to a vector space U, and if $\underset{\sim}{u}_0$ is the zero element in U, then the set

$$\{\underset{\sim}{v} : \underset{\sim}{v} \in V, L(\underset{\sim}{v}) = \underset{\sim}{u}_0\}$$

Definition 1
* * *

is called the kernel* of the morphism. (Another name which is in common use for this set is the null space.) We shall denote the kernel by the letter K.

Notation 1
* * *

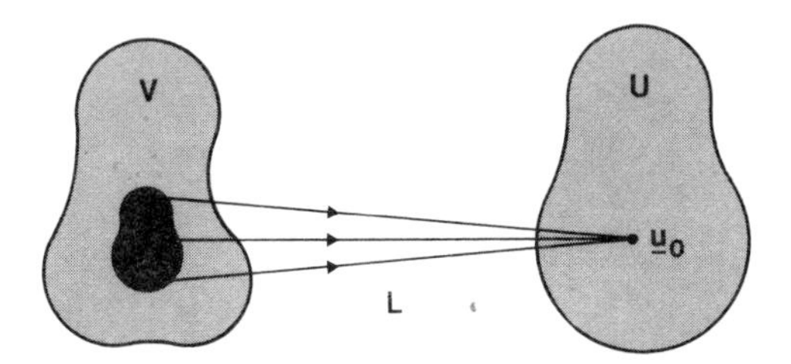

There is one important point to notice here. In *Unit 22, Linear Algebra I*, we defined a *basis* of a vector space V to be a linearly independent set of vectors in V which spans V. We defined the *dimension* of V to be the number of elements in a basis. Now the kernel of a morphism may only contain the zero element of V; i.e. we may have $K = \{\underset{\sim}{0}\}$.

We write $\underset{\sim}{0}$ instead of $\underset{\sim}{v}_0$ here, because you may like to refer to our discussion of the vector space $\{\underset{\sim}{0}\}$ in section 22.2.2 of *Unit 22*. There, we proved that

$$\alpha\underset{\sim}{0} = \underset{\sim}{0},$$

where α is any real number. This means that $\{\underset{\sim}{0}\}$ is a *linearly dependent set*; that is, $\{\underset{\sim}{0}\}$ *does not possess a basis.* We adopt the following definition.

Definition 2
* * *

The dimension of the zero vector space, $\{\underset{\sim}{0}\}$, is zero.

The kernel has some quite remarkable properties. We have already hinted at two of them which are printed in red below.

(1) The kernel itself is a vector space.

We have shown that $L(V)$ is itself a vector space, and in Examples 1 and 2 we have seen that

(2) (dimension of $L(V)$) = (dimension of V) — (dimension of kernel)

The proof of (2) is given in the Appendix at the back of this text. The proof of (1) is not difficult; we have set it as an exercise.

* This is in accordance with the ordinary use of *kernel* for the *nucleus* or *core* of a structure.

Exercise 2

Exercise 2
(2 minutes)

Find the kernel of each of the following morphisms:

(i) $T:(x_1, x_2, x_3) \longmapsto (x_1, x_2, 0)$
where T maps R^3 to R^3.

(ii) $T:(x_1, x_2) \longmapsto (x_1 + x_2, x_1 - 2x_2)$
where T maps R^2 to R^2.

In each case find the dimension of the kernel and verify statement 2 in the text. ■

Exercise 3

Exercise 3
(4 minutes)

Let L be a morphism from a vector space V to a vector space U. Show that the kernel of L is a vector subspace of V.

(HINT: How many of the axioms of a vector space need proving for the kernel? (See section 23.1.2.)) ■

23.1.4 Some Properties of the Kernel

23.1.4

Main Text
* * *

It is remarkable how much we can tell about a morphism just by considering its kernel.

Let L be a morphism of a vector space V to a vector space U, and let K be its kernel. If $\underline{k} \in K$ and $\underline{v} \in V$, then

$$\begin{aligned} L(\underline{v} + \underline{k}) &= L(\underline{v}) + L(\underline{k}) && (L \text{ is a morphism}) \\ &= L(\underline{v}) + \underline{u}_0 && (\text{definition of } K) \\ &= L(\underline{v}) && (\text{axiom 4 for } U), \end{aligned}$$

where $\underline{u}_0$ is the zero element in U. So $\underline{v}$ and $\underline{v} + \underline{k}$, where $\underline{k}$ is any element of the kernel, have the same image.
Suppose now that we want to find *all* the elements in V which map to a given element $\underline{u} \in U$, and that we know one such element $\underline{v}$, i.e.

$$L(\underline{v}) = \underline{u}$$

Then we know immediately that $\underline{v} + \underline{k}$, for all $\underline{k} \in K$, are such elements, and the remarkable thing is that they are in fact *all* the elements which map to $\underline{u}$. We can prove this as follows. Suppose $\underline{v}_1 \in V$ maps to $\underline{u}$, i.e. $L(\underline{v}_1) = \underline{u}$. Then consider $\underline{v}_1 + (-1)\underline{v}$. We have

$$\begin{aligned} L(\underline{v}_1 + (-1)\underline{v}) &= L(\underline{v}_1) + (-1)L(\underline{v}) && (L \text{ is a morphism}) \\ &= \underline{u} + (-1)\underline{u} && (\text{by hypothesis}) \\ &= \underline{u}_0 && (\text{axiom 6 for } U). \end{aligned}$$

But the kernel K contains all those elements which map to $\underline{u}_0$, so

$$\underline{v}_1 + (-1)\underline{v} = \underline{k}_1$$

for some $\underline{k}_1 \in K$. By axiom 1 for V, $\underline{k}_1$ is *unique*. Adding $\underline{v}$ to both sides, we get

$$\underline{v}_1 = \underline{v} + \underline{k}_1,$$

and so $\underline{v}_1$ is of the form $\underline{v}$ + some element of the kernel. This result has important consequences: for instance, if the kernel contains a finite number of elements, n say, then we know that *exactly* n elements of V map to *any* given element of $L(V)$. In particular, if the kernel contains just one element (which will necessarily be the zero element in V), then we know immediately that L is one-one, i.e. an isomorphism. The following examples apply this discussion.

Example 1

Example 1

Apply the above ideas to finding the solution of the equations

$$2x + 3y - z = 1$$
$$x + y - z = 2$$

in terms of vector spaces. ■

Solution of Example 1

One way of expressing the problem is to say that we want to find the set of triples (x, y, z) which satisfy these equations. If L is the mapping from R^3 to R^2 defined by

$$L:(x, y, z) \longmapsto (2x + 3y - z, x + y - z),$$

then we want to find the set which maps to $(1, 2)$.

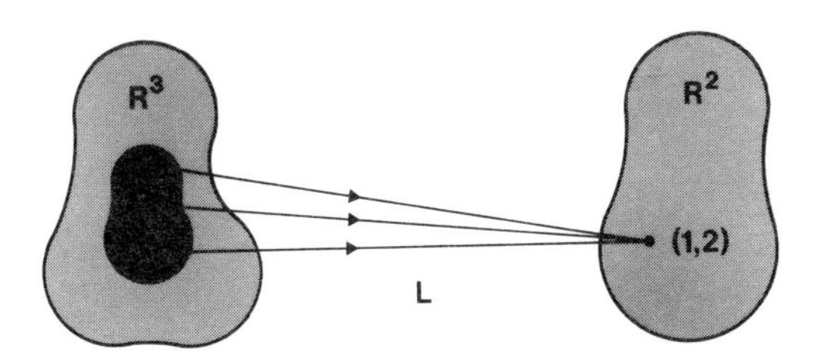

We have seen that we can describe the set which maps to any particular element when we know the kernel and one element of the set. In the context of this example, this means that, if we can find one solution to the equations, we shall be able to find *all* the solutions, simply by adding to that solution each element of the kernel. So we have to find *one solution* and we have to find the *kernel.*

One Solution

If we give x or y or z a particular value, then the equations will be reduced to two equations in two unknowns, which we can solve easily.

For example, if we put $z = 0$, then we obtain the two equations

$$2x + 3y = 1$$
$$x + y = 2,$$

which we can solve to give

$$x = 5, \qquad y = -3.$$

So one solution of the original equations

$$2x + 3y - z = 1$$
$$x + y - z = 2$$

is

$$x = 5, y = -3, z = 0.$$

(continued on page 21)

Solution 23.1.3.2

Solution 23.1.3.2

(i) $\{(0, 0, x_3): x_3 \in R\}$. Any element in this set is a scalar multiple of $(0, 0, 1)$.

(ii) $\{(x_1, x_2): x_1 + x_2 = 0 \text{ and } x_1 - 2x_2 = 0\}$

The pair of simultaneous equations

$$x_1 + x_2 = 0$$

$$x_1 - 2x_2 = 0$$

has the single solution $x_1 = 0, x_2 = 0$. Thus the kernel is the set $\{(0, 0)\}$.

The dimensions of the kernels are as follows:

(i) 1. The dimension of the domain is 3, the dimension of its image set is 2, and $3 - 2 = 1$.

(ii) 0. The dimension of both the domain and its image set is 2. Note that, by defining the dimension of $\{\underset{\sim}{0}\}$ to be zero, we have ensured that (2) is satisfied when $K = \{\underset{\sim}{0}\}$. ■

Solution 23.1.3.3

Solution 23.1.3.3

We need only prove that the elements of the kernel satisfy axioms 1, 4 and 5. As usual, we denote the kernel by K.

Axiom 1 If $\underset{\sim}{k}_1$ and $\underset{\sim}{k}_2$ belong to K, we want to prove that $\underset{\sim}{k}_1 + \underset{\sim}{k}_2 \in K$. We recognize elements of K by the fact that under L they map to $\underset{\sim}{u}_0$, the zero element in U.

$$\begin{aligned} L(\underset{\sim}{k}_1 + \underset{\sim}{k}_2) &= L(\underset{\sim}{k}_1) + L(\underset{\sim}{k}_2) && (L \text{ is a morphism}) \\ &= \underset{\sim}{u}_0 + \underset{\sim}{u}_0 && (\underset{\sim}{k}_1, \underset{\sim}{k}_2 \in K) \\ &= \underset{\sim}{u}_0 && (\text{axiom 4 for } U) \end{aligned}$$

Therefore, $\underset{\sim}{k}_1 + \underset{\sim}{k}_2 \in K$.

Axiom 4 We have already shown (Exercise 23.1.2.2) that $L(\underset{\sim}{v}_0) = \underset{\sim}{u}_0$. Therefore $\underset{\sim}{v}_0 \in K$.

Axiom 5 Let $\underset{\sim}{k} \in K$, then

$$\begin{aligned} L(\alpha\underset{\sim}{k}) &= \alpha L(\underset{\sim}{k}) && (L \text{ is a morphism}) \\ &= \alpha\underset{\sim}{u}_0 && (\underset{\sim}{k} \in K) \\ &= \underset{\sim}{u}_0 \end{aligned}$$

This last step has been proved in *Unit 22, Linear Algebra I*, p. 40. The verification of axioms 1 and 5 can be amalgamated by using the definition of a morphism in the form given in Equation 23.1.2.3. ■

(contined from page 19)

The Kernel

The kernel, K, is the set of triples (x, y, z) which map to $(0, 0)$, i.e. which satisfy the equations

$$2x + 3y - z = 0$$

$$x + y - z = 0$$

Just as before, we can solve these equations by giving x or y or z a particular value and then trying to solve the resulting two equations in two unknowns; but this time it is not much help, because we simply get the *one* solution, and we want *all* solutions. But if, for example, we give z a general value and put $z = k$, then these equations become

$$2x + 3y = k$$

$$x + y = k,$$

which we can solve to give

$$x = 2k, \qquad y = -k.$$

So one element of the kernel is $(2k, -k, k)$, and by varying k we get all the elements. So K is the set

$$\{(2k, -k, k) : k \in R\}.$$

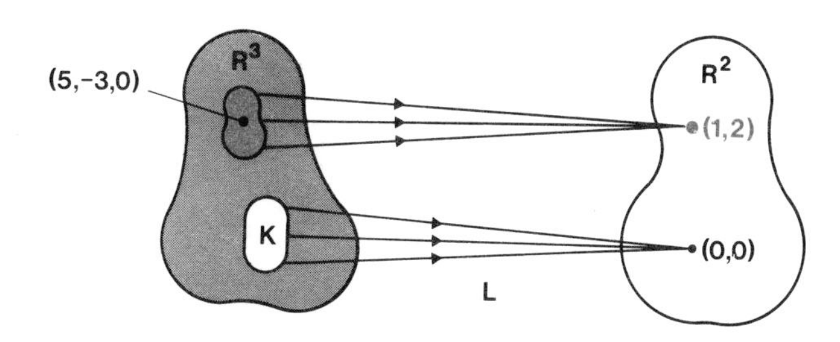

We want to find the red shaded set in the above diagram.
We know one element — how do we find the others?

Any solution of the original equations

$$2x + 3y - z = 1$$

$$x + y - z = 2$$

is obtained by adding an element of the kernel to $(5, -3, 0)$.

So the complete solution set is

$$\{(5 + 2k, -3 - k, k) : k \in R\},$$

and the theory we have developed assures us that this set contains *all* the possible solutions to the original equations.

(Check that these *are* solutions by substituting into the original equations.)
Notice that we can solve related problems like

$$2x + 3y - z = 7$$

$$x + y - z = 99,$$

where the right-hand sides of the equations are changed, very quickly — all we need to do is to find one particular solution and then add on each of the elements of the (same) kernel. We shall discuss problems of this type in considerable detail in *Unit 26, Linear Algebra III*. ■

Exercise 1

Exercise 1
(3 minutes)

By putting $x = 0$, find a particular solution of the equations

$$2x + 3y - z = 2$$

$$x + y + z = 1$$

Find the solution set of the equations. ■

Exercise 2

Exercise 2
(3 minutes)

We can map the vector space P_n, of all polynomial functions of degree *n or less*, to itself by the differentiation operator:

$$D: p \longmapsto p' \qquad (p \in P_n).$$

We have already seen that this mapping is a morphism. What is the kernel? What significance does this have in integration? ■

Discussion
* *

Problems such as the following often arise in mathematics. We are given a mapping, M say, from a set A to a set B, and are required to find all the elements x of A such that

$$M: x \longmapsto b,$$

where b is a given element of B. That is to say, we have to solve the equation $M(x) = b$. Sometimes there is no solution (if b does not belong to $M(A)$); sometimes there is a unique solution; sometimes there are many solutions. In a very wide class of problems, A and B are vector spaces and M is a morphism. We have just seen that in these cases we can (in principle) adopt a standard procedure. We first find the kernel, K, that is, the solution set of

$$M(\underset{\sim}{x}) = \underset{\sim}{0}.$$

(This is usually a much easier problem than the original one.) We then find any one particular solution of

$$M(\underset{\sim}{x}) = \underset{\sim}{b},$$

and then combine this with each element of K, to get the complete solution set.

We shall apply these techniques later in the course, and we shall also meet the idea of a kernel in contexts other than that of a vector space.

23.1.5 An Algebra of Morphisms

23.1.5

Main Text

We shall now use some of our earlier work on functions. In *Unit 1, Functions* we saw various ways in which functions could be combined. In particular, we saw that there are two distinct types of combination: composition, or combination by successive performance, and the "arithmetic" combinations which can be "induced" from operations defined in the codomain. In *Unit 1*, we gave particular examples of this latter type. We saw that when the codomain is a set of real numbers we can use the operations of addition, multiplication and so on, to define addition of functions, multiplication of functions, and so on.

If T_1 and T_2 are functions mapping a vector space V to a vector space U, then we have two vector space operations defined on U which can be used to define new functions of V to U:

Definition 1

$$T_1 + T_2 : \underset{\sim}{v} \longmapsto T_1(\underset{\sim}{v}) + T_2(\underset{\sim}{v}) \qquad (v \in V)$$

$$\alpha T_1 : \underset{\sim}{v} \longmapsto \alpha(T_1(\underset{\sim}{v})) \qquad (v \in V)$$

where α is a real number.

Exercise 1

Exercise 1
(3 minutes)

Show that, if T_1 and T_2 are morphisms, then $T_1 + T_2$ and αT_1 are both morphisms. (If you find that you need to look up the solutions to this exercise, then do so, but the solutions will have little value if you only read them, because they are cluttered with symbols. So once you have read them, try writing out at least one part on your own.) ■

Exercise 2

Exercise 2
(4 minutes)

Given two vector spaces V and U, we can consider the set of all morphisms from V to U. It is a remarkable fact that this set of functions with the operations we have just defined is itself a vector space. Prove this statement.

(You will need to check each of the axioms of a vector space. They are listed on page 11.) ■

Main Text

Finally, we shall show* that, if T_1 is a morphism of the vector space V to the vector space U, and T_2 is a morphism of U to the vector space W, then $T_2 \circ T_1$ is a morphism of V to W. With the usual notation, we have

$$\begin{aligned} T_2 \circ T_1(\alpha_1 \underset{\sim}{v}_1 + \alpha_2 \underset{\sim}{v}_2) &= T_2(T_1(\alpha_1 \underset{\sim}{v}_1 + \alpha_2 \underset{\sim}{v}_2)) \\ &= T_2(\alpha_1 T_1(\underset{\sim}{v}_1) + \alpha_2 T_1(\underset{\sim}{v}_2)) \\ &\qquad \text{(since } T_1 \text{ is a morphism)} \\ &= \alpha_1 T_2(T_1(\underset{\sim}{v}_1)) + \alpha_2 T_2(T_1(\underset{\sim}{v}_2)) \\ &\qquad \text{(since } T_2 \text{ is a morphism)} \\ &= \alpha_1 T_2 \circ T_1(\underset{\sim}{v}_1) + \alpha_2 T_2 \circ T_1(\underset{\sim}{v}_2) \end{aligned}$$

i.e. Equation 23.1.2.3 holds, so $T_2 \circ T_1$ is a morphism.

* What we prove here is a special case of the general result that the composition of two morphisms is a morphism. The only reason we prove it here is because we have not proved it before in the course: we could have proved it earlier in *Unit 3, Operations and Morphisms*.

Solution 23.1.4.1

Solution 23.1.4.1

Putting $x = 0$ in both equations, gives

$$3y - z = 2$$

$$y + z = 1,$$

which have the solution

$$y = \tfrac{3}{4}, z = \tfrac{1}{4}$$

Thus one solution is

$$x = 0, y = \tfrac{3}{4}, z = \tfrac{1}{4}.$$

To find the kernel, we have to solve the equations

$$2x + 3y - z = 0$$

$$x + y + z = 0$$

If we put $x = k$, we get

$$3y - z = -2k$$

$$y + z = -k,$$

which have the solution

$$y = -\tfrac{3}{4}k, z = -\tfrac{1}{4}k;$$

so the kernel is the set

$$\{(k, -\tfrac{3}{4}k, -\tfrac{1}{4}k) : k \in R\}.$$

The complete solution is therefore

$$\{(k, \tfrac{3}{4} - \tfrac{3}{4}k, \tfrac{1}{4} - \tfrac{1}{4}k) : k \in R\}.$$

(You may recall from *Unit 15, Differentiation II* that the equations $2x + 3y - z = 2$ and $x + y + z = 1$ are equations of planes. The set of points

$$\{(k, \tfrac{3}{4} - \tfrac{3}{4}k, \tfrac{1}{4} - \tfrac{1}{4}k) : k \in R\}$$

corresponds to the line formed by the intersection of these two planes.) ■

Solution 23.1.4.2

Solution 23.1.4.2

The kernel is the set of all polynomial functions which map to the zero function $(x \longmapsto 0 \ (x \in R))$; that is, the set of all constant functions,

$$\{f : x \longmapsto k \ (x \in R), k \in R\}.$$

The problem of integration is that of solving equations of the form

$$D(p) = f,$$

where f is given.

Since the kernel contains an infinite number of elements, the integration problem has an infinite number of solutions. If p is one solution, then the set of all solutions is

$$\{p + f : f : x \longmapsto k \ (k \in R), k \in R\}.$$ ■

Solution 1

Solution 1

Let $T = T_1 + T_2$.

$$\begin{aligned} T(\underline{u} + \underline{v}) &= (T_1 + T_2)(\underline{u} + \underline{v}) \\ &= T_1(\underline{u} + \underline{v}) + T_2(\underline{u} + \underline{v}) && \text{(definition of } T_1 + T_2) \\ &= T_1(\underline{u}) + T_1(\underline{v}) + T_2(\underline{u}) + T_2(\underline{v}) \\ & && (T_1 \text{ and } T_2 \text{ are morphisms}) \\ &= (T_1(\underline{u}) + T_2(\underline{u})) + (T_1(\underline{v}) + T_2(\underline{v})) && \text{(axioms 2 and 3)} \\ &= (T_1 + T_2)(\underline{u}) + (T_1 + T_2)(\underline{v}) && \text{(definition of } T_1 + T_2) \\ &= T(\underline{u}) + T(\underline{v}) \end{aligned}$$

Thus the first condition for a morphism is satisfied. Now let α be any real number.

$$\begin{aligned} T(\alpha\underline{u}) &= (T_1 + T_2)(\alpha\underline{u}) \\ &= T_1(\alpha\underline{u}) + T_2(\alpha\underline{u}) && \text{(definition of } T_1 + T_2) \\ &= \alpha T_1(\underline{u}) + \alpha T_2(\underline{u}) && (T_1 \text{ and } T_2 \text{ are morphisms}) \\ &= \alpha(T_1 + T_2)(\underline{u}) && \text{(axiom 7 and definition of } T_1 + T_2) \\ &= \alpha T(\underline{u}) \end{aligned}$$

Thus the second condition is also satisfied: T is a morphism. (The proof could be shortened by using the definition of a morphism given by Equation 23.1.2.3, and showing that

$$T(\alpha_1\underline{u}_1 + \alpha_2\underline{u}_2) = \alpha_1 T(\underline{u}_1) + \alpha_2 T(\underline{u}_2).)$$

We also have to prove that αT_1 is a morphism. We use Equation 23.1.2.3.

$$\begin{aligned} \alpha T_1(\alpha_1\underline{u}_1 + \alpha_2\underline{u}_2) &= \alpha(T_1(\alpha_1\underline{u}_1 + \alpha_2\underline{u}_2)) && \text{(by definition of } \alpha T_1) \\ &= \alpha(\alpha_1 T_1(\underline{u}_1) + \alpha_2 T_1(\underline{u}_2)) && (T_1 \text{ is a morphism}) \\ &= (\alpha\alpha_1)T_1(\underline{u}_1) + (\alpha\alpha_2)T_1(\underline{u}_2) && \text{(axioms 7 and 9)} \\ &= \alpha_1(\alpha T_1)(\underline{u}_1) + \alpha_2(\alpha T_1)(u_2) && \text{(axiom 9)} \end{aligned}$$

as required. ■

Solution 2

Solution 2

Closure under addition and multiplication by a scalar have been proved in Exercise 1 (axioms 1 and 5).

The zero mapping

$O : \underline{v} \longmapsto \underline{u}_0 \qquad (\underline{v} \in V)$

belongs to the set because it is a morphism of V to U (axiom 4).

The other axioms may be investigated directly from the definition of addition of functions and the definition of multiplication of a function by a scalar.

It is important in this proof to be very clear about the equality of two functions; two functions are *equal* if they have the same domain and the same image for each element of the domain. ■

23.1.6 Summary

We have concentrated our attention on those mappings of vector spaces which are morphisms. Under a morphism the image of a vector space is itself a vector space. A particularly important subset of the domain of such morphisms is the set of elements which map to the zero vector in the codomain. We call this set of elements the *kernel* of the morphism, and it has some remarkable properties. It is itself a vector space; its dimension tells us the difference between the dimensions of the domain and the image space; it tells us whether the morphism is many-one or one-one; it provides us with a method of solving equations such as $L(\underline{v}) = \underline{u}$, where L is a morphism of a vector space. If L is a homomorphism then it is, by definition, a many-one mapping, and so we can expect this equation to have more than one solution.

We saw that if we know the kernel of the morphism and if we know one solution of the equation $L(\underline{v}) = \underline{u}$, then we can easily generate every other solution by using the elements of the kernel. We shall see later in this course how this method has direct application to the solution of problems in the theory of linear equations, in complex numbers and in differential equations, and how an analogous method can be developed in group theory. We also showed that, given two vector spaces V and U, the set of all morphisms from V to U is itself a vector space. Finally, we showed that the composition of two (suitably defined) morphisms of vector spaces is itself a morphism.

These results are listed below.

Let L be a morphism from a vector space V to a vector space U. Then,

(i) $L(V)$ is a vector subspace of U;
(ii) $K = \{\underline{k} \in V : L(\underline{k}) = \underline{u}_0\}$ is a vector subspace of V, called the *kernel* of the morphism;
(iii) (dimension of V) = (dimension of $L(V)$) + (dimension of K);
(iv) if $\underline{v}_1$ is a solution of $L(\underline{v}) = (\underline{u})$, then the set of all solutions of this equation is $\{\underline{v}_1 + \underline{k} : \underline{k} \in K\}$;
(v) $K = \{\underline{v}_0\} \Leftrightarrow L$ is an isomorphism;
(vi) the set of all morphisms from V to U is a vector space;
(vii) if L_1 is a morphism from U to a vector space W, then $L_1 \circ L$ is a morphism from V to W.

23.2 MATRICES

23.2

23.2.0 Introduction

23.2.0

Introduction
* *

In this second part of the text we introduce a new idea — that of a *matrix*. A *matrix* is simply a rectangular (or square) array of numbers such as you might see on a bus timetable or on the board outside a cinema, showing the starting times of the films. Such arrays of numbers arise when we try to quantify almost any investigation.

For example, the following two arrays might represent a survey of the voting inclinations in two constituencies.

	Party X	Party Y	Party Z	Don't Know
Men	30%	35%	10%	25%
Women	25%	50%	15%	10%

	Party X	Party Y	Party Z	Don't Know
Men	50%	40%	5%	5%
Women	40%	40%	2%	18%

There are various ways of combining arrays. In this example, adding corresponding entries and dividing by 2 to give

	Party X	Party Y	Party Z	Don't Know
Men	40%	$37\frac{1}{2}$%	$7\frac{1}{2}$%	15%
Women	$32\frac{1}{2}$%	45%	$8\frac{1}{2}$%	14%

might be a meaningful thing to do, to give a sort of "average" voting inclination.

In the first section of this second part of the text we show how morphisms between vector spaces can be represented by matrices. This correspondence between mappings and matrices leads us to define some ways of combining matrices. We may want to define special combinations for special applications, but the ones we define here are the most widely used.

(There is a lot of algebraic and numerical manipulation in section 23.2. If you are unfamiliar with matrices, then it will not help you much just to read through the calculations. We advise you to work through all the calculations as you come to them, and just use the manipulations given in the text as a check for your own.)

23.2.1 Linear Equations

23.2.1

Discussion
* *

In the first part of this text we discussed several examples of mappings from R^3 to R^2 or R^3 to R^3, and we saw in section 23.1.4 that these mappings are connected with simultaneous equations. It is fairly clear that three equations, each expressing one of x_1', x_2' and x_3' in terms of x_1, x_2 and x_3, will tell us how to map (x_1, x_2, x_3) to (x_1', x_2', x_3'). If these equations hold for all real values of x_1, x_2, x_3, then they will define a mapping from R^3 to R^3. For example, the three equations

Equations (1)

$$x_1' = \sin x_1$$
$$x_2' = \sin x_2$$
$$x_3' = \sin x_3$$

define such a mapping, as do the equations

$$x_1' = x_1 + x_2 + x_3$$
$$x_2' = 2x_1 - x_2$$
$$x_3' = 2x_2 + 3x_3$$

Equations (2)

where the right-hand sides are linear combinations of x_1, x_2 and x_3.

Exercise 1

Exercise 1
(2 minutes)

Do Equations (1) define a morphism from R^3 to R^3?

Do Equations (2) define a morphism from R^3 to R^3? ■

Discussion
* *

It seems that *any* set of three equations of the form:

$$x_1' = a_1x_1 + a_2x_2 + a_3x_3$$
$$x_2' = b_1x_1 + b_2x_2 + b_3x_3$$
$$x_3' = c_1x_1 + c_2x_2 + c_3x_3$$

Equations (3)

where the a's, b's and c's are real numbers, defines a morphism from R^3 to R^3. A set of equations of this form is called a set of linear equations, and our suggestion can be put in more general terms as in the following proposition.

Definition 1
* *

Theorem
* * *

Any set of m linear equations, each expressing one of m variables

$$x_1', x_2', \ldots, x_m' \text{ in terms of } n \text{ variables } x_1, x_2, \ldots, x_n$$

defines a morphism of R^n to R^m.

The proposition is easily confirmed, for if the equations are

$$x_1' = a_1x_1 + a_2x_2 + \cdots + a_nx_n$$
$$x_2' = b_1x_1 + b_2x_2 + \cdots + b_nx_n$$
$$\cdots\cdots\cdots\cdots$$
$$\cdots\cdots\cdots\cdots$$
$$x_m' = q_1x_1 + q_2x_2 + \cdots + q_nx_n$$

Equations (4)

they certainly define a *mapping* T of R^n to R^m:

$$T:(x_1, x_2, \ldots, x_n) \longmapsto (x_1', x_2', \ldots, x_m').$$

For the variables $x_1, x_2, \ldots, x_n$ can each take any real value, and so $(x_1, x_2, \ldots, x_n)$ can be any element of R^n. We have to prove that T is a morphism, i.e. that

$$T(\alpha x_1, \alpha x_2, \ldots, \alpha x_n) = \alpha T(x_1, x_2, \ldots, x_n)$$

where α is any real number, and

$$T(x_1 + y_1, x_2 + y_2, \ldots, x_n + y_n)$$
$$= T(x_1, x_2, \ldots, x_n) + T(y_1, y_2, \ldots, y_n)$$

Exercise 2

Exercise 2
(2 minutes)

Prove that T is a morphism by proving the last two statements in the text. ■

Discussion
* *

We thus have the result that every set of linear equations of the form of Equations (4) defines a *morphism*.

This in itself would not be so interesting, if it were not for the fact that we can represent *any* morphism from R^n to R^m by such a set of equations. We can do this as follows.

Discussion
* *

We take as a basis for R^n the set of n vectors:

$$\underset{\sim}{e}_1 = (1, 0, 0, 0, \ldots, 0),$$
$$\underset{\sim}{e}_2 = (0, 1, 0, 0, \ldots, 0),$$
$$\underset{\sim}{e}_3 = (0, 0, 1, 0, \ldots, 0),$$
$$\ldots$$
$$\underset{\sim}{e}_n = (0, 0, 0, 0, \ldots, 1).$$

An arbitrary element of R^n can be written:

$$\underset{\sim}{x} = (x_1, x_2, x_3, \ldots, x_n) = x_1\underset{\sim}{e}_1 + x_2\underset{\sim}{e}_2 + x_3\underset{\sim}{e}_3 + \cdots + x_n\underset{\sim}{e}_n.$$

If T is a morphism from R^n to R^m, then

$$T(\underset{\sim}{x}) = T(x_1\underset{\sim}{e}_1 + x_2\underset{\sim}{e}_2 + \cdots + x_n\underset{\sim}{e}_n) = x_1 T(\underset{\sim}{e}_1) + \cdots + x_n T(\underset{\sim}{e}_n).$$

We see that the base vectors tell us the whole story. Each of the base vectors has an image in R^m; suppose

$$T(\underset{\sim}{e}_1) = (a_1, b_1, \ldots, q_1)$$
$$T(\underset{\sim}{e}_2) = (a_2, b_2, \ldots, q_2)$$
$$\ldots$$
$$T(\underset{\sim}{e}_n) = (a_n, b_n, \ldots, q_n)$$

Then

$$\begin{aligned} T(\underset{\sim}{x}) = \; & x_1(a_1, b_1, \ldots, q_1) \\ & + x_2(a_2, b_2, \ldots, q_2) \\ & \ldots \\ & + x_n(a_n, b_n, \ldots, q_n) \\ = \; & (a_1x_1 + a_2x_2 + \cdots + a_nx_n, b_1x_1 + \cdots + b_nx_n, \ldots). \end{aligned}$$

If

$$T(\underset{\sim}{x}) = T(x_1, x_2, \ldots, x_n) = (x'_1, x'_2, \ldots, x'_m),$$

then

$$T(\underset{\sim}{x}) = x'_1\underset{\sim}{e}_1 + x'_2\underset{\sim}{e}_2 + \cdots + x'_m\underset{\sim}{e}_m.$$

(The basis $\{\underset{\sim}{e}_1, \underset{\sim}{e}_2, \ldots, \underset{\sim}{e}_m\}$ for R^m is shown on page 41, written in terms of column vectors, instead of row vectors as we have here.)

We have

Equations (5)

$$\begin{aligned} x'_1 &= a_1x_1 + a_2x_2 + \cdots + a_nx_n \\ x'_2 &= b_1x_1 + b_2x_2 + \cdots + b_nx_n \\ &\ldots \\ x'_m &= q_1x_1 + q_2x_2 + \cdots + q_nx_n \end{aligned}$$

Thus, the morphism can be represented by Equations (5).

Theorem
* * *

There is a one-one correspondence between sets of linear equations and morphisms from R^n to R^m. Any set of linear equations of the form (5) defines a morphism from R^n to R^m, and any morphism from R^n to R^m defines a set of linear equations of the form (5).

Discussion
*

We mentioned earlier that a morphism between vector spaces is often called a *linear transformation*. We now see why — the *transformation* refers to the mapping part, and *linear* refers to the properties we have just discussed. It is from here also that we get the term *Linear Algebra*.

In *Unit 26, Linear Algebra III*, we shall take up the topic of sets of linear equations in much more detail. In the meantime we shall use the ideas we have just been discussing to introduce the topic of *matrices*, which is of great importance in *Linear Algebra* and mathematics generally, and which we shall use when we discuss linear equations in detail.

Solution 1

Solution 1

Let $\underset{\sim}{x} = (x_1, x_2, x_3)$ and $\underset{\sim}{y} = (y_1, y_2, y_3)$.

For Equations (1) we have the mapping S, where

$$S(\underset{\sim}{x}) = (\sin x_1, \sin x_2, \sin x_3)$$

$$S(\underset{\sim}{y}) = (\sin y_1, \sin y_2, \sin y_3)$$

$$S(\underset{\sim}{x} + \underset{\sim}{y}) = (\sin (x_1 + y_1), \sin (x_2 + y_2), \sin (x_3 + y_3))$$

and

$$S(\underset{\sim}{x}) + S(\underset{\sim}{y}) = (\sin x_1 + \sin y_1, \sin x_2 + \sin y_2, \sin x_3 + \sin y_3)$$

and so

$$S(\underset{\sim}{x} + \underset{\sim}{y}) \neq S(\underset{\sim}{x}) + S(\underset{\sim}{y}).$$

For Equations (2) we have the mapping T, where

$$T(\underset{\sim}{x}) = (x_1 + x_2 + x_3, 2x_1 - x_2, 2x_2 + 3x_3)$$

and

$$\begin{aligned} T(\underset{\sim}{x} + \underset{\sim}{y}) = \big(&x_1 + y_1 + x_2 + y_2 + x_3 + y_3, \\ &2(x_1 + y_1) - (x_2 + y_2), \\ &2(x_2 + y_2) + 3(x_3 + y_3)\big) \\ = {}& T(\underset{\sim}{x}) + T(\underset{\sim}{y}) \end{aligned}$$

Similarly

$$T(\alpha \underset{\sim}{x}) = \alpha T(\underset{\sim}{x})$$

where α is any real number.
Thus Equations (2) define a morphism, but Equations (1) do not. ■

Solution 2

Solution 2

The first statement follows from the fact that, for example,

$$a_1(\alpha x_1) + a_2(\alpha x_2) + \cdots + a_n(\alpha x_n)$$
$$= \alpha(a_1 x_1 + a_2 x_2 + \cdots + a_n x_n).$$

a process which can be applied to each of the equations.

Similarly, we have, for example,

$$a_1(x_1 + y_1) + a_2(x_2 + y_2) + \cdots + a_n(x_n + y_n)$$
$$(a_1 x_1 + a_2 x_2 + \cdots + a_n x_n) + (a_1 y_1 + a_2 y_2 + \cdots + a_n y_n).$$

The second statement follows from an application of this process to each component of $T(x_1 + y_1, \ldots, x_n + y_n)$. ■

23.2.2 Arrays of Numbers

23.2.2

Notation
* * *

We have seen that Equations 23.2.1.5 define a mapping from R^n to R^m (given bases for R^n and R^m). Now $(x_1, x_2, \ldots, x_n)$ and $(x'_1, x'_2, \ldots, x'_m)$ are simply arbitrary elements in the respective vector spaces; the equations themselves are adequately specified by the array of numbers

$$\begin{matrix} a_1 & a_2 & \cdots & a_n \\ b_1 & b_2 & \cdots & b_n \\ & & \cdots & \\ & & \cdots & \\ q_1 & q_2 & \cdots & q_n. \end{matrix}$$

Such an array of numbers is called a matrix.

Definition 1
* * *

Often we want to talk about this array as an entity in itself, as opposed to a collection of elements. To do this we either put parentheses round the array or refer to it by a single letter. Sometimes we do both, thus we write

$$A \text{ or } \underbrace{\begin{pmatrix} a_1 & a_2 & \cdots & a_n \\ b_1 & b_2 & \cdots & b_n \\ \cdots & & & \\ q_1 & q_2 & \cdots & q_n \end{pmatrix}}_{n \text{ columns}} \begin{matrix} \leftarrow \\ \leftarrow \\ \\ \leftarrow \end{matrix} \Bigg\} \; m \text{ rows}$$

Thus, a matrix with m rows and n columns defines a mapping from R^n to R^m (the bases being given or understood), and the mapping is generally referred to by the name of the matrix.

Example 1

Example 1

(i) The mapping of R^2 to R^2 defined by

$$x' = 2x + 3y$$

$$y' = 3x - y$$

is represented by the matrix

$$\begin{pmatrix} 2 & 3 \\ 3 & -1 \end{pmatrix}$$

(ii) The matrix

$$\begin{pmatrix} 1 & 2 & 3 \\ 4 & 0 & 6 \end{pmatrix}$$

represents the mapping of R^3 to R^2 defined by

$$x_1' = x_1 + 2x_2 + 3x_3$$

$$x_2' = 4x_1 + 0x_2 + 6x_3$$

■

Definition 2
* * *

Two matrices are said to be equal if they represent the same mapping (with respect to bases which are given or understood). Thus matrices A and B are equal if they have the same number of rows and columns (they must represent mappings with the same domain and codomain), and the corresponding elements are equal. If A and B are equal, we write $A = B$. For example,

$$\begin{pmatrix} 1 & 2 \\ 3 & 4 \end{pmatrix} = \begin{pmatrix} 1 & 2 \\ 3 & 4 \end{pmatrix}$$

$$\begin{pmatrix} 1 & 2 & 3 & 4 \\ 5 & 6 & 7 & 8 \end{pmatrix} = \begin{pmatrix} 1 & 2 & 3 & 4 \\ 5 & 6 & 7 & 8 \end{pmatrix}$$

$$\begin{pmatrix} 1 & 2 \\ 3 & 4 \end{pmatrix} \neq \begin{pmatrix} 2 & 4 \\ 6 & 8 \end{pmatrix}$$

$$\begin{pmatrix} 1 & 2 & 3 & 4 \\ 0 & 0 & 0 & 0 \end{pmatrix} \neq \begin{pmatrix} 1 & 2 & 3 & 4 \\ 5 & 6 & 7 & 8 \end{pmatrix}$$

$$\begin{pmatrix} 1 & 2 & 3 & 4 \end{pmatrix} \neq \begin{pmatrix} 1 & 2 & 3 & 4 \\ 0 & 0 & 0 & 0 \end{pmatrix}$$

Discussion
* *

Since a matrix is an alternative way of specifying a morphism, we should be able to tell something about the morphism just by looking at the matrix. Investigating the properties of matrices is equivalent to investigating morphisms between vector spaces.

23.2.3 Combining Matrices

23.2.3

Discussion
* *

We can *represent* a set of equations by a matrix, but we can also go one better and *rewrite* the equations in terms of this matrix. Consider, for example, the equations

$$x'_1 = 3x_1 + 2x_2 + 1x_3$$
$$x'_2 = 1x_1 + 1x_2 + 3x_3$$

Equations (1)

They specify a mapping from R^3 to R^2 under which

$$(x_1, x_2, x_3) \longmapsto (x'_1, x'_2),$$

and are represented by the matrix

$$A = \begin{pmatrix} 3 & 2 & 1 \\ 1 & 1 & 3 \end{pmatrix}$$

We can write the equations in the following way:

$$(x'_1, x'_2) = A(x_1, x_2, x_3),$$

but there are plenty of other ways in which we could write them. We shall choose a way which leads to results consistent with the existing literature on matrices. We write

$$(x'_1, x'_2) \text{ as } \begin{pmatrix} x'_1 \\ x'_2 \end{pmatrix}$$

and

$$(x_1, x_2, x_3) \text{ as } \begin{pmatrix} x_1 \\ x_2 \\ x_3 \end{pmatrix};$$

that is, we write the elements of R^2 and R^3 as arrays (or lists) of numbers, rather than in co-ordinate form. These lists are just special kinds of matrices, matrices with *one* column. So we can rewrite the equations in matrix notation as

$$\begin{pmatrix} x'_1 \\ x'_2 \end{pmatrix} = A \,\square \begin{pmatrix} x_1 \\ x_2 \\ x_3 \end{pmatrix},$$

Equation (2)

where $\square$ is a combination of matrices which we have to define to make Equation (2) equivalent to Equations (1).

From our definition of equality of matrices, $A \,\square \begin{pmatrix} x_1 \\ x_2 \\ x_3 \end{pmatrix}$ must be a matrix with *one* column and *two* rows, and the first element must be x'_1 and the second x'_2. But x'_1 and x'_2 are given by Equations (1), so

$$A \,\square \begin{pmatrix} x_1 \\ x_2 \\ x_3 \end{pmatrix} = \begin{pmatrix} 3x_1 + 2x_2 + x_3 \\ x_1 + x_2 + 3x_3 \end{pmatrix}$$

Thus

$$\begin{pmatrix} 3 & 2 & 1 \\ 1 & 1 & 3 \end{pmatrix} \square \begin{pmatrix} x_1 \\ x_2 \\ x_3 \end{pmatrix} = \begin{pmatrix} 3x_1 + 2x_2 + 1x_3 \\ 1x_1 + 1x_2 + 3x_3 \end{pmatrix}$$

and for a general mapping from R^3 to R^2:

$$\begin{pmatrix} a & b & c \\ d & e & f \end{pmatrix} \Box \begin{pmatrix} \alpha \\ \beta \\ \gamma \end{pmatrix} = \begin{pmatrix} a\alpha + b\beta + c\gamma \\ d\alpha + e\beta + f\gamma \end{pmatrix}$$

The *first element* in the third matrix comes from the *first row* of the first matrix, and is obtained by multiplying each element of that row with the corresponding element in the second matrix and adding together the products thus formed. The *second element* comes from the *second row* of the first matrix by applying the same procedure.

Similarly,

$$\begin{pmatrix} 1 & 2 & 3 \\ -2 & 3 & -4 \end{pmatrix} \Box \begin{pmatrix} 1 \\ -2 \\ 3 \end{pmatrix} = \begin{pmatrix} 1 \times 1 + 2 \times (-2) + 3 \times 3 \\ (-2) \times 1 + 3 \times (-2) + (-4) \times 3 \end{pmatrix}$$

$$= \begin{pmatrix} 6 \\ -20 \end{pmatrix}$$

We can extend this operation to deal with longer lists and matrices with more rows and columns. This extension corresponds to manipulating equations specifying mappings from R^n to R^m, where the column corresponding to an element of R^n has n elements

$$\begin{pmatrix} x_1 \\ x_2 \\ \cdot \\ \vdots \\ x_n \end{pmatrix}, \text{ and the left-hand matrix is } \left.\begin{pmatrix} a_1 & a_2 & \cdots & a_n \\ b_1 & b_2 & \cdots & b_n \\ \cdots & & & \\ \vdots & & & \\ q_1 & q_2 & \cdots & q_n \end{pmatrix}\right\} m \text{ rows}$$

This matrix must, of course, have the same number of columns as there are elements in the list matrix corresponding to an element of R^n, because of the form of the linear equations.

Example 1

Example 1

(i) $$\begin{pmatrix} 1 & 2 & 3 \\ 4 & 5 & 6 \\ 7 & 8 & 9 \end{pmatrix} \Box \begin{pmatrix} 1 \\ -1 \\ 2 \end{pmatrix} = \begin{pmatrix} 1 \times 1 + 2 \times (-1) + 3 \times 2 \\ 4 \times 1 + 5 \times (-1) + 6 \times 2 \\ 7 \times 1 + 8 \times (-1) + 9 \times 2 \end{pmatrix} = \begin{pmatrix} 5 \\ 11 \\ 17 \end{pmatrix}$$

(ii) $$\begin{pmatrix} 1 & 2 \\ 3 & 4 \\ 5 & 6 \end{pmatrix} \Box \begin{pmatrix} 7 \\ 8 \end{pmatrix} = \begin{pmatrix} 23 \\ 53 \\ 83 \end{pmatrix}$$

(iii) The set of equations

$$2x_1 + 3x_2 = 2$$
$$1x_1 - 2x_2 = 3$$
$$3x_1 + 1x_2 = 4$$

can be written in the form

$$\begin{pmatrix} 2 & 3 \\ 1 & -2 \\ 3 & 1 \end{pmatrix} \Box \begin{pmatrix} x_1 \\ x_2 \end{pmatrix} = \begin{pmatrix} 2 \\ 3 \\ 4 \end{pmatrix}$$

■

Exercise 1

Exercise 1
(2 minutes)

Calculate

(i) $\begin{pmatrix} 1 & -1 \\ 2 & 0 \\ 3 & 4 \end{pmatrix} \square \begin{pmatrix} 3 \\ -2 \end{pmatrix}$ (ii) $\begin{pmatrix} 1 & 2 & 3 \\ -1 & 0 & 4 \end{pmatrix} \square \begin{pmatrix} 1 \\ 2 \\ 3 \end{pmatrix}$ ■

Discussion
* *

At the end of section 23.1.5, we proved that the composition of two morphisms, their sum, and the scalar multiple of a morphism are all morphisms. It follows that we can define combinations of matrices corresponding to each of these combinations of morphisms. Here we shall consider the combination which corresponds to composition. We shall consider the other cases in section 23.2.5.

Consider, for example, the morphisms T_1 and T_2 of R^2 to R^2 defined by

$$T_1 : (x_1, x_2) \longmapsto (1x_1 + 1x_2, 2x_1 + 1x_2)$$
$$T_2 : (x_1, x_2) \longmapsto (1x_1 + 2x_2, 3x_1 + 2x_2)$$

Under the composite mapping $T_2 \circ T_1$ we have

$$T_2 \circ T_1 : (x_1, x_2) \longmapsto T_2(T_1(x_1, x_2))$$

and

$$\begin{aligned} T_2(T_1(x_1, x_2)) &= T_2(1x_1 + 1x_2, 2x_1 + 1x_2) \\ &= (1(x_1 + x_2) + 2(2x_1 + x_2), 3(x_1 + x_2) \\ &\quad + 2(2x_1 + x_2)) \\ &= (5x_1 + 3x_2, 7x_1 + 5x_2) \end{aligned}$$

Writing

$$T_1(x_1, x_2) = (x_1', x_2') \quad \text{and} \quad T_2(x_1', x_2') = (x_1'', x_2''),$$

we have:

$$x_1' = 1x_1 + 1x_2$$
$$x_2' = 2x_1 + 1x_2$$

Equations (3)

and

$$x_1'' = 1x_1' + 2x_2'$$
$$x_2'' = 3x_1' + 2x_2'$$

Equations (4)

Substituting from Equations (3) into Equations (4), we get

$$x_1'' = 5x_1 + 3x_2$$
$$x_2'' = 7x_1 + 5x_2$$

Equations (5)

as before.

In terms of matrices, the matrix representation for T_1 is

$$A_1 = \begin{pmatrix} 1 & 1 \\ 2 & 1 \end{pmatrix}$$

(see Equations (3)), and the matrix representation for T_2 is

$$A_2 = \begin{pmatrix} 1 & 2 \\ 3 & 2 \end{pmatrix}$$

(see Equations (4)). And we have shown that the matrix representation for $T_2 \circ T_1$ is

$$A_3 = \begin{pmatrix} 5 & 3 \\ 7 & 5 \end{pmatrix}$$

(see Equations (5)).

If we use the symbol $*$ to represent the operation of combining matrices which corresponds to the composition of mappings, we have

$$A_2 * A_1 = A_3$$

i.e.

$$\begin{pmatrix} 1 & 2 \\ 3 & 2 \end{pmatrix} * \begin{pmatrix} 1 & 1 \\ 2 & 1 \end{pmatrix} = \begin{pmatrix} 5 & 3 \\ 7 & 5 \end{pmatrix}$$

Equation (6)

We now know how to combine A_1 and A_2 in this case, but how can we work out $B * A$ for *any* matrices B and A? Indeed, does $B * A$ *make sense* for *any* matrices B and A?

First of all, how do we calculate $B * A$? Looking at Equation (6), we see that the element in the *first* row and *first* column of A_3 is 5, and

$$5 = 1 \times 1 + 2 \times 2,$$

and this expression is obtained by multiplying the elements in the *first* row of A_2 by the corresponding elements in the *first* column of A_1 and adding together the products.

$$\begin{pmatrix} 1 & 2 \\ & \end{pmatrix} * \begin{pmatrix} 1 & \\ 2 & \end{pmatrix} = \begin{pmatrix} 5 & \\ & \end{pmatrix}$$

Combining rows of A_2 with columns of A_1 in this way, we get the other elements of A_1.

$$\begin{pmatrix} & \\ 3 & 2 \end{pmatrix} * \begin{pmatrix} 1 & \\ 2 & \end{pmatrix} = \begin{pmatrix} & \\ 7 & \end{pmatrix}$$

$$\begin{pmatrix} 1 & 2 \\ & \end{pmatrix} * \begin{pmatrix} & 1 \\ & 1 \end{pmatrix} = \begin{pmatrix} & 3 \\ & \end{pmatrix}$$

$$\begin{pmatrix} & \\ 3 & 2 \end{pmatrix} * \begin{pmatrix} & 1 \\ & 1 \end{pmatrix} = \begin{pmatrix} & \\ & 5 \end{pmatrix}$$

In the general case of matrices with two rows and two columns, we define $*$ by

$$\begin{pmatrix} e & f \\ g & h \end{pmatrix} * \begin{pmatrix} a & b \\ c & d \end{pmatrix} = \begin{pmatrix} ea + fc & eb + fd \\ ga + hc & gb + hd \end{pmatrix}$$

Definition 1
* * *

Exercise 2

Exercise 2
(2 minutes)

If

$$\begin{aligned} x_1' &= ax_1 + bx_2 \\ x_2' &= cx_1 + dx_2 \end{aligned} \qquad \text{(i)}$$

and

$$\begin{aligned} x_1'' &= ex_1' + fx_2' \\ x_2'' &= gx_1' + hx_2' \end{aligned} \qquad \text{(ii)}$$

find x_1'' and x_2'' in terms of x_1 and x_2, and compare your answer with the definition of $*$. ■

(*continued on page 36*)

Solution 1

Solution 1

(i) $\begin{pmatrix} 5 \\ 6 \\ 1 \end{pmatrix}$, because $1 \times 3 + (-1) \times (-2) = 5$
and $2 \times 3 + 0 \times (-2) = 6$
and $3 \times 3 + 4 \times (-2) = 1$

(ii) $\begin{pmatrix} 14 \\ 11 \end{pmatrix}$, because $1 \times 1 + 2 \times 2 + 3 \times 3 = 14$
and $(-1) \times 1 + 0 \times 2 + 4 \times 3 = 11$ ■

Solution 2

Solution 2

Substituting from (i) into (ii), we get

$$x_1'' = e(ax_1 + bx_2) + f(cx_1 + dx_2)$$

$$x_2'' = g(ax_1 + bx_2) + h(cx_1 + dx_2)$$

i.e.

$$x_1'' = (ea + fc)x_1 + (eb + fd)x_2$$

$$x_2 = (ga + hc)x_1 + (gb + hd)x_2$$

and the matrix representing these equations is

$$\begin{pmatrix} ea + fc & eb + fd \\ ga + hc & gb + hd \end{pmatrix}$$ ■

(*continued from page 35*)

Discussion * *

We can extend the definition of * to more general matrices. So far we have combined matrices with two rows and two columns — corresponding to combining a mapping from R^2 to R^2 with another mapping from R^2 to R^2. It makes sense to combine mappings other than these; a morphism T_1 from R^m to R^p, say, can be followed by a morphism T_2 from R^p to R^n, say.

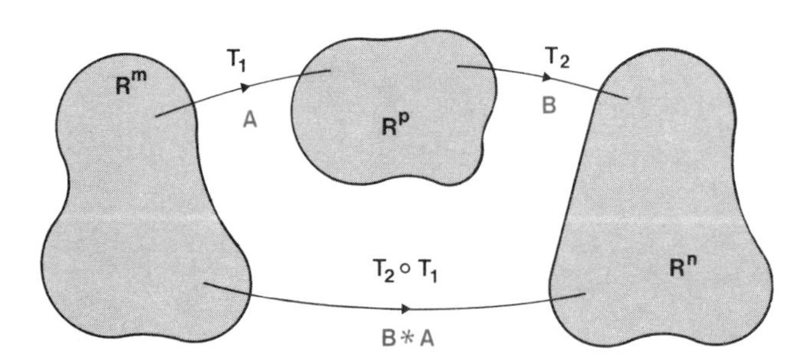

We know that T_1 will be represented by a matrix with m columns and p rows, and T_2 will be represented by a matrix with p columns and n rows. So we can extend the definition of * to form $B * A$ for all sorts of matrices *provided* the *number of columns in B* is the same as the *number of rows in A*. Since $T_2 \circ T_1$ is a morphism from R^m to R^n, the combined matrix will have n rows and m columns, and in general $B * A$ will be a matrix with the *same number of rows as B* and the *same number of columns as A*.

$$\underset{\text{rows}}{n}\overset{p}{\begin{pmatrix}\text{columns} \\ \\ \\ \end{pmatrix}} * \underset{\text{rows}}{p}\overset{m}{\begin{pmatrix}\text{columns} \\ \\ \\ \end{pmatrix}} = \underset{\text{rows}}{n}\overset{m}{\begin{pmatrix}\text{columns} \\ \\ \\ \end{pmatrix}}$$

We shall not follow through all the details, but illustrate how we might proceed in a simple example.

Example 2

Example 2

$$\text{If } B = \begin{pmatrix} 1 & 2 \\ 3 & 4 \\ 5 & 6 \end{pmatrix} \quad \text{and} \quad A = \begin{pmatrix} 1 \\ 2 \end{pmatrix},$$

how do we calculate $B * A$? A represents the mapping from R^1 to R^2 specified by the equations

Equations (7)

$$x_1' = 1x_1$$
$$x_2' = 2x_1$$

and B represents the mapping from R^2 to R^3 specified by the equations

Equations (8)

$$x_1'' = 1x_1' + 2x_2'$$
$$x_2'' = 3x_1' + 4x_2'$$
$$x_3'' = 5x_1' + 6x_2'$$

Substituting Equations (7) into Equations (8), we find that the composite mapping has matrix

$$B * A = \begin{pmatrix} 5 \\ 11 \\ 17 \end{pmatrix}$$

This is a matrix with three rows and one column; we expect this because the result of mapping R^1 to R^2 (matrix A) and then R^2 to R^3 (matrix B) is a mapping from R^1 to R^3 (matrix $B * A$).

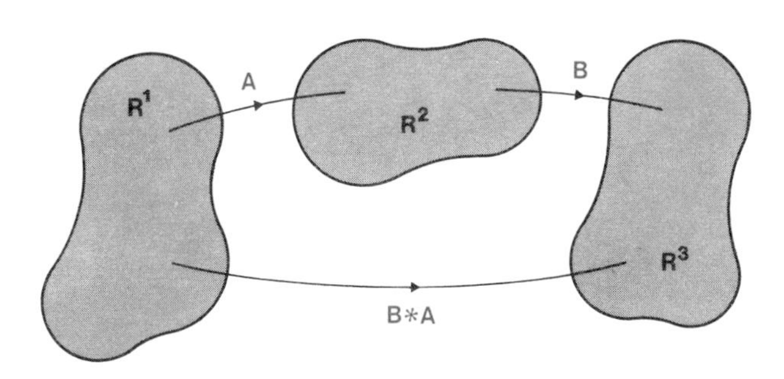

The elements in the matrix $B * A$ are calculated by exactly the same method as before: the element 5 comes from the *first* row of B and the *first* (and only) column of A: $5 = 1 \times 1 + 2 \times 2$. The element 11 comes from the *second* row of B and the *first* (and only) column of A: $11 = 3 \times 1 + 4 \times 2$. And $17 = 5 \times 1 + 6 \times 2$. ■

If you look back at page 32 where we discussed the operation □, you will see that □ is just a special case of $*$. We can define $*$ for general matrices (to correspond to the composition of morphisms) as follows.

Definition 2
* * *

The combination $B * A$ is defined for any matrices B and A, provided that the *number of columns in B* is the same as the *number of rows in A*. If

$$B * A = C,$$

then the element in row i and column j of C is obtained from row i of B and column j of A; it is calculated by adding together the termwise products of the elements in this row and column.

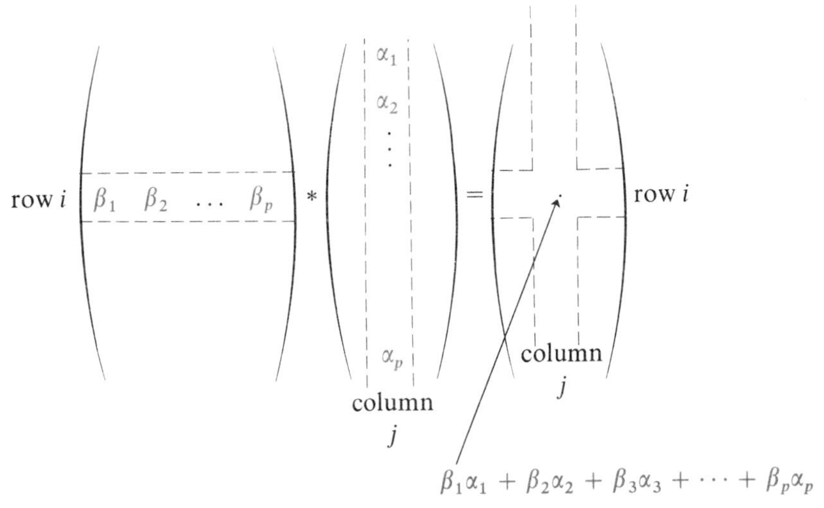

If B has m rows and p columns and A has p rows and n columns, then $C = B * A$ will have m rows and n columns.

Example 3

Example 3

(Make sure you can obtain these results for yourself.)

(i) $$\begin{pmatrix} 1 & -2 \\ -1 & 3 \\ 2 & 4 \end{pmatrix} * \begin{pmatrix} 1 & 3 & -4 \\ -1 & -2 & 1 \end{pmatrix} = \begin{pmatrix} 3 & 7 & -6 \\ -4 & -9 & 7 \\ -2 & -2 & -4 \end{pmatrix}$$

(ii) $$(1 \quad 2 \quad 3 \quad 4 \quad 5) * \begin{pmatrix} 1 \\ 1 \\ 1 \\ 1 \\ 1 \end{pmatrix} = (15)$$

(iii) $$\begin{pmatrix} 1 & -2 \\ -1 & 3 \\ 2 & 4 \end{pmatrix} * \begin{pmatrix} 1 & 3 & -4 \\ -1 & -2 & 1 \\ 1 & 2 & 3 \end{pmatrix}$$ is not defined, because

the number of columns in the left-hand matrix is *not* the same as the number of rows in the right-hand matrix. ■

Exercise 3

Exercise 3
(3 minutes)

(i) Calculate $\begin{pmatrix} 1 & 2 \\ -1 & 3 \end{pmatrix} * \begin{pmatrix} 3 & -1 \\ 2 & 1 \end{pmatrix}$

(ii) Calculate $\begin{pmatrix} 1 & 3 & 5 \\ -1 & 2 & -1 \end{pmatrix} * \begin{pmatrix} 1 & 1 \\ 2 & 2 \\ 0 & -1 \end{pmatrix}$

(iii) Is $*$ (a) commutative?
(b) associative? ■

23.2.4 Some Special Matrices

23.2.4

Rotation Matrices

Discussion
* *

In the first part of this text we had several examples of morphisms specified by sets of equations, and we discussed their geometric interpretation. For example, we saw that the equations

$$\begin{aligned} x_1' &= -x_2 = 0x_1 + (-1)x_2 \\ x_2' &= x_1 = 1x_1 + 0x_2 \end{aligned}$$

can be regarded as specifying a mapping which rotates the plane through an angle $\frac{\pi}{2}$ anti-clockwise about the origin. In terms of matrices, we can say that $\begin{pmatrix} 0 & -1 \\ 1 & 0 \end{pmatrix}$ represents a rotation of the plane through $\frac{\pi}{2}$ anti-clockwise.

Suppose we start the other way round and look for matrices to do some specified geometric job.

Example 1

Example 1

Consider a mapping which rotates the plane about the origin through an angle θ anti-clockwise.

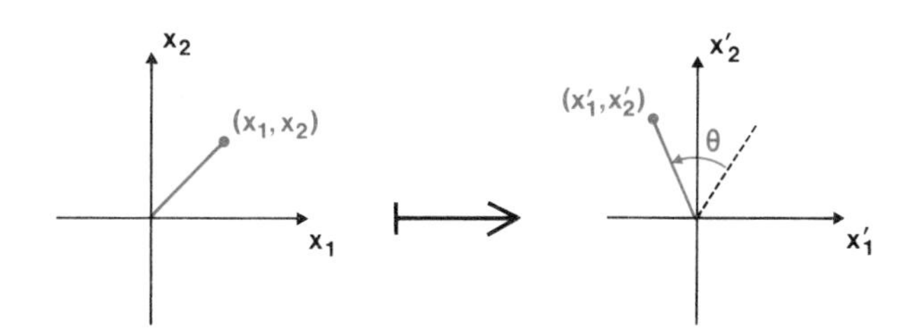

If we call the corresponding matrix R_θ, then

$$\begin{pmatrix} x_1' \\ x_2' \end{pmatrix} = R_\theta \begin{pmatrix} x_1 \\ x_2 \end{pmatrix}$$

where x_1, x_2, x_1' and x_2' are related as in the diagram. We can find R_θ easily by considering the basis $\left\{ \begin{pmatrix} 1 \\ 0 \end{pmatrix}, \begin{pmatrix} 0 \\ 1 \end{pmatrix} \right\}$ for R^2.

A little trigonometry tells us that

(See RB2)

$$\begin{pmatrix} 1 \\ 0 \end{pmatrix} \longmapsto \begin{pmatrix} \cos\theta \\ \sin\theta \end{pmatrix}$$

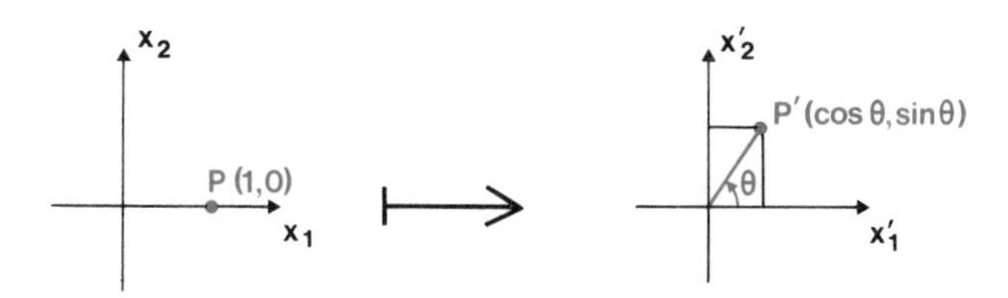

(continued on page 40)

Solution 23.2.3.3

Solution 23.2.3.3

(i) $\begin{pmatrix} 7 & 1 \\ 3 & 4 \end{pmatrix}$

(ii) $\begin{pmatrix} 7 & 2 \\ 3 & 4 \end{pmatrix}$

(iii) (a) It is easy enough to find a counter-example to show that $*$ is not commutative. For example,

$$\begin{pmatrix} 1 & 0 \\ 0 & 0 \end{pmatrix} * \begin{pmatrix} 0 & 0 \\ 1 & 0 \end{pmatrix} = \begin{pmatrix} 0 & 0 \\ 0 & 0 \end{pmatrix}$$

and

$$\begin{pmatrix} 0 & 0 \\ 1 & 0 \end{pmatrix} * \begin{pmatrix} 1 & 0 \\ 0 & 0 \end{pmatrix} = \begin{pmatrix} 0 & 0 \\ 1 & 0 \end{pmatrix}$$

(b) This is more difficult, because $*$ is in fact, associative.

A direct proof for $*$ could be tricky, but we can go about it another way. Remembering that $*$ for matrices corresponds to composition of morphisms, which are just special kinds of functions, we can show that the composition of functions (where it is possible) is associative. Suppose A, B, C, D are sets and f, g, h are functions such that

$f: A \longmapsto B, g: B \longmapsto C, h: C \longmapsto D$

then if $a \in A$,

$$\begin{aligned} ((h \circ g) \circ f)(a) &= (h \circ g)(f(a)) \\ &= h(g(f(a))) \\ &= h((g \circ f)(a)) \\ &= (h \circ (g \circ f))(a) \end{aligned}$$

Thus, composition of functions is associative, which implies that $*$ is also associative. ■

(*continued from page 39*)

and $\begin{pmatrix} 0 \\ 1 \end{pmatrix} \longmapsto \begin{pmatrix} -\sin\theta \\ \cos\theta \end{pmatrix}$

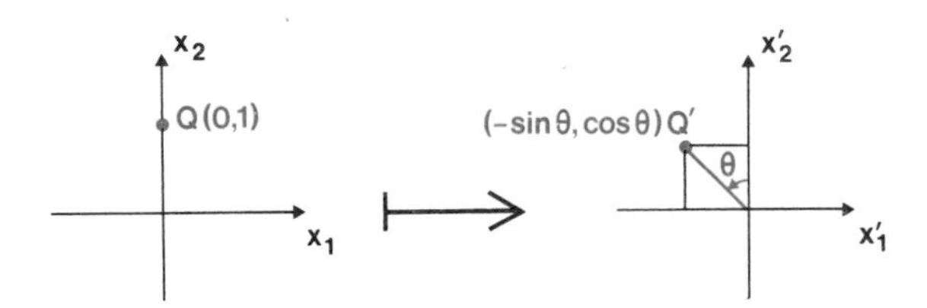

R_θ has two rows and two columns because the mapping is one-one and the image set of R^2 is R^2. Thus if

$$R_\theta = \begin{pmatrix} a & b \\ c & d \end{pmatrix},$$

we know that

$$\begin{pmatrix} a & b \\ c & d \end{pmatrix} * \begin{pmatrix} 1 \\ 0 \end{pmatrix} = \begin{pmatrix} \cos\theta \\ \sin\theta \end{pmatrix}$$

and

$$\begin{pmatrix} a & b \\ c & d \end{pmatrix} * \begin{pmatrix} 0 \\ 1 \end{pmatrix} = \begin{pmatrix} -\sin\theta \\ \cos\theta \end{pmatrix}$$

Since

$$\begin{pmatrix} a & b \\ c & d \end{pmatrix} * \begin{pmatrix} 1 \\ 0 \end{pmatrix} = \begin{pmatrix} a \\ c \end{pmatrix}$$

and

$$\begin{pmatrix} a & b \\ c & d \end{pmatrix} * \begin{pmatrix} 0 \\ 1 \end{pmatrix} = \begin{pmatrix} b \\ d \end{pmatrix}$$

we find that

$$a = \cos\theta, \quad c = \sin\theta, \quad b = -\sin\theta \text{ and } d = \cos\theta.$$

Thus

$$R_\theta = \begin{pmatrix} \cos\theta & -\sin\theta \\ \sin\theta & \cos\theta \end{pmatrix}$$

■

Rotation Matrix
* *
Generalization
*

You may have noticed in this example that the columns of R_θ are precisely the images of the base vectors $\begin{pmatrix} 1 \\ 0 \end{pmatrix}$ and $\begin{pmatrix} 0 \\ 1 \end{pmatrix}$ under the mapping. This is not special to this example, for it is easily seen that if A is the matrix of a morphism T from R^m, then the columns of A are precisely

$$T\underset{\sim}{e}_1, T\underset{\sim}{e}_2, T\underset{\sim}{e}_3, \ldots, T\underset{\sim}{e}_m,$$

where $\underset{\sim}{e}_1, \underset{\sim}{e}_2, \underset{\sim}{e}_3, \ldots, \underset{\sim}{e}_m$ are the particular base vectors

$$\underset{\sim}{e}_1 = \begin{pmatrix} 1 \\ 0 \\ 0 \\ \vdots \\ 0 \end{pmatrix}, \underset{\sim}{e}_2 = \begin{pmatrix} 0 \\ 1 \\ 0 \\ \vdots \\ 0 \end{pmatrix}, \ldots, \underset{\sim}{e}_m = \begin{pmatrix} 0 \\ 0 \\ 0 \\ \vdots \\ 1 \end{pmatrix}$$

(Compare this with the discussion on page 29.)

The subject of Linear Algebra is full of elegant yet simple results like this.

Exercise 1

Exercise 1
(1 minute)

Find the matrix which corresponds to a "stretching of the plane", so that the distance of every point from the origin is multiplied by a number k. ■

Exercise 2

Exercise 2
(1 minute)

Find the matrix which corresponds to the identity mapping, which leaves every point in the plane unchanged. ■

The Identity Matrix

Under the operation of composition of functions, the identity function leaves every function unchanged. For example, the function:

$$f : x \longmapsto x \qquad (x \in R)$$

is such that $f \circ g = g \circ f = g$ for any function g with domain and codomain R. This property has its counterpart in matrix algebra. If we write $I = \begin{pmatrix} 1 & 0 \\ 0 & 1 \end{pmatrix}$, then $I * A = A * I = A$ for *any* matrix with two rows and

(*continued on page 42*)

Solution 1

Solution 1

Since $\begin{pmatrix}1\\0\end{pmatrix} \longmapsto \begin{pmatrix}k\\0\end{pmatrix}$ and $\begin{pmatrix}0\\1\end{pmatrix} \longmapsto \begin{pmatrix}0\\k\end{pmatrix}$, the matrix required is $\begin{pmatrix}k & 0\\0 & k\end{pmatrix}$. ■

Solution 2

Solution 2

The matrix is $\begin{pmatrix}1 & 0\\0 & 1\end{pmatrix}$.

(The special case of Solution 1 when $k = 1$.) ■

(continued from page 41)

two columns. Indeed, if we write $I_{m,m}$ for the matrix with m rows and m columns, such that every entry is zero except for 1's as the diagonal entries from the top left-hand corner to the bottom right-hand corner, then if A is any matrix with m rows and n columns, we have

$$I_{m,m} * A = A * I_{n,n} = A$$

Usually, the suffices are dropped, and we just write I for the identity matrix appropriate to the particular situation.

Identity Matrix
* * *

Exercise 3 (Optional)

Exercise 3
(3 minutes)

You may find the following interesting.

(i) What is the result of rotating the plane anti-clockwise about the origin through an angle α and then through an angle β?

(ii) Complete the following equation:

$$R_\beta * R_\alpha \boxed{}$$

(iii) Complete the following equation

$$R_{\beta+\alpha} = \begin{pmatrix}\cos(\beta+\alpha) & \boxed{}\\ \boxed{} & \boxed{}\end{pmatrix}$$

(iv) Write down R_α and R_β and calculate $R_\beta * R_\alpha$ directly.

(v) Compare your answers to (iii) and (iv). Can you deduce anything from this comparison? ■

23.2.5 More Ways of Combining Matrices

23.2.5

Discussion ★★

The operation $*$ arose from our interpretation of the composition of morphisms in terms of matrices, and was made possible because we had established the correspondence between morphisms and matrices.

There is an interesting result concerning morphisms which we obtained towards the end of section 23.1.5, and which we now want to take up again. We showed that the set of all morphisms from a given vector space V to a given vector space U *itself* forms a vector space. We know that, for example, any morphism from R^m to R^n is represented by a matrix with n rows and m columns, and conversely, that any such matrix represents a morphism from R^m to R^n. So the set of all matrices with n rows and m columns forms a vector space (for given n and m) for operations corresponding to those for morphisms, i.e. addition of functions and multiplication of a function by a number. It is fairly easy to see that the corresponding matrix operations are as follows:

Definition 1 ★★★

(i) If k is any real number, then kA is the matrix obtained by multiplying each element of A by k.

Definition 2 ★★★

(ii) If A and B are two matrices each with n rows and m columns, then $A \triangle B$ is the matrix with n rows and m columns obtained by adding corresponding elements of A and B.

Example 1

Example 1

(i) $$3\begin{pmatrix} 1 & 2 \\ -1 & 3 \\ 4 & 7 \end{pmatrix} = \begin{pmatrix} 3 & 6 \\ -3 & 9 \\ 12 & 21 \end{pmatrix}$$

(ii) $$\begin{pmatrix} 1 & 2 \\ -1 & 3 \\ 4 & 7 \end{pmatrix} \triangle \begin{pmatrix} 3 & 6 \\ -3 & 9 \\ 12 & 21 \end{pmatrix} = \begin{pmatrix} 1+3 & 2+6 \\ (-1)+(-3) & 3+9 \\ 4+12 & 7+21 \end{pmatrix}$$

$$= \begin{pmatrix} 4 & 8 \\ -4 & 12 \\ 16 & 28 \end{pmatrix}$$

(iii) $$\begin{pmatrix} 1 & 2 \\ -1 & 3 \end{pmatrix} \triangle \begin{pmatrix} 3 & -3 & 12 \\ 6 & 9 & 21 \end{pmatrix}$$

is not defined: it is equivalent to trying to "add" two functions with different domains. ■

Discussion ★★

It is clear from the definition of $\triangle$ that it is both commutative and associative. Because of its obvious similarities to addition, we use the symbol $+$ for $\triangle$; for example, we write

$$\begin{pmatrix} 2 & 3 \\ 1 & 4 \end{pmatrix} + \begin{pmatrix} 2 & 7 \\ -2 & -4 \end{pmatrix} = \begin{pmatrix} 4 & 10 \\ -1 & 0 \end{pmatrix}$$

Definition 3 ★★★

and we call the operation matrix addition.

Definition 4 ★★★

We can use these two operations to define a third operation — matrix subtraction — by

$$A - B = A + (-1)B$$

The matrix $(-1)B$ is usually written as $-B$.

So now we have three operations: $*$, multiplication by a scalar, and $+$. The two operations $+$ and $*$ are binary operations on sets of matrices, although we have to take some care over defining the sets in which they operate. We have noted some of their properties: $+$ is commutative and

(*continued on page 44*)

Solution 23.2.4.3

Solution 23.2.4.3

If we perform two rotations such as these successively, the result is a rotation through an angle $\beta + \alpha$, which is of course the same as a rotation through $\alpha + \beta$. Hence

(i) A rotation through $\beta + \alpha$.

(ii) $R_\beta * R_\alpha = R_{\beta+\alpha}$

(iii)
$$R_{\beta+\alpha} = \begin{pmatrix} \cos(\beta+\alpha) & -\sin(\beta+\alpha) \\ \sin(\beta+\alpha) & \cos(\beta+\alpha) \end{pmatrix}$$

(iv)
$$R_\beta * R_\alpha = \begin{pmatrix} \cos\beta & -\sin\beta \\ \sin\beta & \cos\beta \end{pmatrix} * \begin{pmatrix} \cos\alpha & -\sin\alpha \\ \sin\alpha & \cos\alpha \end{pmatrix}$$
$$= \begin{pmatrix} \cos\beta\cos\alpha - \sin\beta\sin\alpha & -\cos\beta\sin\alpha - \sin\beta\cos\alpha \\ \sin\beta\cos\alpha + \cos\beta\sin\alpha & -\sin\beta\sin\alpha + \cos\beta\cos\alpha \end{pmatrix}$$

(v) Comparing (iii) and (iv), we obtain the formulas

$$\cos(\beta+\alpha) = \cos\beta\cos\alpha - \sin\beta\sin\alpha$$

$$\sin(\beta+\alpha) = \sin\beta\cos\alpha + \cos\beta\sin\alpha$$ ■

(*continued from page 43*)

associative; $*$ is associative but not commutative. We have not as yet seen how $*$ and $+$ interact. In particular, are they distributive over each other?

Exercise 1

Exercise 1
(2 minutes)

By choosing some simple matrices, prove that $+$ is not distributive over $*$, i.e. that for some suitable* matrices A, B and C

$$A + (B * C) \neq (A + B) * (A + C)$$ ■

Discussion
* *

It is a relatively simple matter to disprove a conjecture by a counter-example, as in Exercise 1, but to *prove* a conjecture we have to argue in general terms. We shall not do this here, because the proof does not illustrate anything very important (except, perhaps, a need for an improved notation, which will be developed later); however, we record the fact that $*$ *is distributive over* $+$. That is, for suitable matrices A, B and C,

$$A * (B + C) = (A * B) + (A * C)$$

and

$$(A + B) * C = (A * C) + (B * C)$$

Notice that we must make *both* statements because $*$ is not commutative.

There is one difficulty that we must mention; it concerns terminology. It is widespread practice to refer to the operation $*$ as *matrix multiplication* and drop the symbol $*$, writing AB instead of $A * B$. This is most unfortunate, for although $*$ is associative, and distributive over $+$, just as for $\times$ and $+$ in the set of real numbers, $*$ is *not* commutative, and whereas $+$ comes from addition of mappings, $*$ comes not from multiplication but from composition.

However, because the practice is widespread we shall refer to the operation as *matrix multiplication* for the remainder of the course, and drop the symbol $*$.

In the next section we collect together some of the differences between matrices and numbers, and also some of the similarities of the two algebras.

* By "suitable" we mean matrices with appropriate numbers of rows and columns, so that the operations are defined.

23.2.6 Matrix Algebra and the Algebra of Numbers

23.2.6

Summary
★★

A major difference between matrices and numbers is that, whereas we can add or multiply any two numbers, this is not the case for matrices. If we restrict ourselves to a set of matrices with the same (fixed) number of rows as columns, then we can add or multiply any two matrices from the set. We take it as implicit, then, in this section that *all the matrices have the same number of rows as columns, and that this number is the same for all the matrices.* (A matrix with the same number of rows as columns is called a *square* matrix.)

Similarities

	Numbers	Matrices
Addition is commutative.	$a + b = b + a$	$A + B = B + A$
Addition is associative.	$(a + b) + c = a + (b + c)$	$(A + B) + C = A + (B + C)$
There is a zero element.	$a + 0 = 0 + a = a$	$A + O = O + A = A$, where O is the matrix in which every entry is zero.
Each element has an additive inverse.	$(-b) + b = 0$	$(-B) + B = O$
Multiplication is associative.	$(ab)c = a(bc)$	$(AB)C = A(BC)$
There is an "identity" element.	$a1 = 1a = a$	$AI = IA = A$, where I is the identity matrix defined on page 42
"Multiplication" is distributive over addition.	$a(b + c) = ab + ac$ $(a + b)c = ac + bc$	$A(B + C) = AB + AC$ $(A + B)C = AC + BC$
Certain product rules hold.	$a0 = 0a = 0$ $a(-b) = -ab$ $(-a)(-b) = ab$	$AO = OA = O$ $A(-B) = -AB$ $(-A)(-B) = AB$

Differences

(i) Number multiplication is commutative; matrix multiplication is not. That is, there exist matrices A, B such that $AB \neq BA$.

As a consequence, the matrix expansion $(A + B)^2 = A^2 + 2AB + B^2$ is *false*, in general. The correct expansion is $(A + B)^2 = A^2 + AB + BA + B^2$. The matrix $AB + BA$ is equal to $2AB$ if and only if A and B commute, i.e. $AB = BA$. This may happen in certain cases (for example if $B = I$), but it is not generally true.

(ii) The product of two non-zero numbers is non-zero, but the product of two non-zero matrices may be equal to the zero matrix.

For example, if

$$A = \begin{pmatrix} 1 & 1 \\ 1 & 1 \end{pmatrix} \text{ and } B = \begin{pmatrix} 1 & 1 \\ -1 & -1 \end{pmatrix},$$

then $AB = O$. (Note also that $BA \neq O$.) This example also illustrates the existence of a non-zero matrix B such that $B^2 = O$.

(iii) The cancellation law holds for numbers, but not for matrices.

That is, if a, b, c are numbers and $a \neq 0$, then

$$(ab = ac) \Rightarrow (b = c)$$

But for matrices,

$$(AB = AC) \not\Rightarrow (B = C).$$

For example, if A and B are as above, then $AB = AO$, but B is not equal to the zero matrix, i.e. A cannot be cancelled.

(continued on page 46)

Solution 23.2.5.1

Solution 23.2.5.1

For example, with

$$A = \begin{pmatrix} 1 & 2 \\ 3 & 4 \end{pmatrix}, B = \begin{pmatrix} 1 & -2 \\ 3 & -1 \end{pmatrix}, C = \begin{pmatrix} 1 & 1 \\ 3 & -2 \end{pmatrix},$$

we have

$$\begin{pmatrix} 1 & 2 \\ 3 & 4 \end{pmatrix} + \left[\begin{pmatrix} 1 & -2 \\ 3 & -1 \end{pmatrix} * \begin{pmatrix} 1 & 1 \\ 3 & -2 \end{pmatrix}\right] = \begin{pmatrix} -4 & 7 \\ 3 & 9 \end{pmatrix}$$

whereas

$$\left[\begin{pmatrix} 1 & 2 \\ 3 & 4 \end{pmatrix} + \begin{pmatrix} 1 & -2 \\ 3 & -1 \end{pmatrix}\right] * \left[\begin{pmatrix} 1 & 2 \\ 3 & 4 \end{pmatrix} + \begin{pmatrix} 1 & 1 \\ 3 & -2 \end{pmatrix}\right]$$

$$= \begin{pmatrix} 4 & 6 \\ 30 & 24 \end{pmatrix}$$ ■

(*continued from page 45*)

(iv) The equation $x^2 = 0$ has the unique solution $x = 0$, but the matrix equation $X^2 = O$ has an infinite number of solutions.

It can be shown that the general solution of the matrix equation $X^2 = O$ for matrices with two rows and columns is

$$X = \begin{pmatrix} r & s \\ -\frac{r^2}{s} & -r \end{pmatrix} \quad \text{for arbitrary } r \text{ and non-zero } s,$$

or

$$X = \begin{pmatrix} 0 & 0 \\ t & 0 \end{pmatrix} \quad \text{for arbitrary } t.$$

(v) The equation $x^2 = -1$ has no solution (in real numbers), but the matrix equation $X^2 = -I$ does have solutions.

We single out two solutions for special attention. These are the matrices

$$C = \begin{pmatrix} 0 & -1 \\ 1 & 0 \end{pmatrix} \text{ and } -C = \begin{pmatrix} 0 & 1 \\ -1 & 0 \end{pmatrix}$$

Note that the matrix $-I$ is equal to the rotation matrix R_π corresponding to a rotation about the origin through the angle π. The matrix C is the rotation matrix $R_{\frac{1}{2}\pi}$. What transformation does $-C$ correspond to?

Exercise 1

Exercise 1
(2 minutes)

Give expansions of $(A - B)^2$ and $(A - B)^3$ which are valid for all square matrices. (There are 8 terms in the second expansion.) ■

Exercise 2

Exercise 2
(2 minutes)

The equation $x^2 = x$ has precisely two solutions, $x = 0$ or 1. By giving an example of a matrix with 2 rows and 2 columns, show that the matrix equation $X^2 = X$ has solutions other than $X = O$ or I. ■

23.3 APPENDIX

The "Dimension" Theorem

This Appendix is optional.

THEOREM

If L is a morphism from a vector space V to a vector space U, then

$$(\text{dimension of } L(V)) + (\text{dimension of kernel}) = (\text{dimension of } V)$$

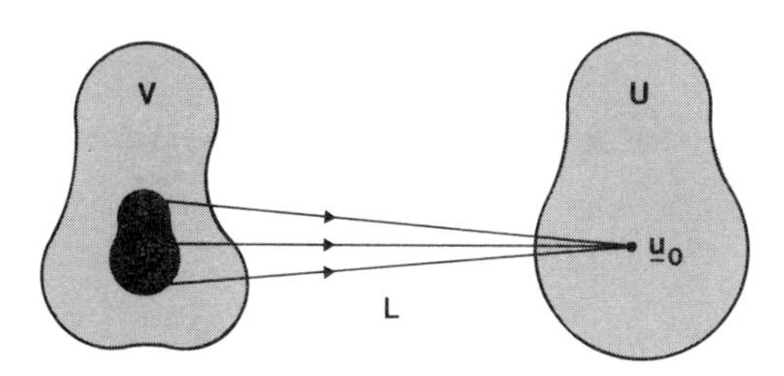

PROOF

Let $\{\underline{u}_1, \underline{u}_2, \underline{u}_3, \ldots, \underline{u}_p\}$ be a basis for $L(V)$ (i.e. let $L(V)$ have dimension p), and let $\{\underline{k}_1, \underline{k}_2, \underline{k}_3, \ldots, \underline{k}_m\}$ be a basis for the kernel (i.e. let the kernel have dimension m). Since each of the u's belongs to $L(V)$, there must be vectors in V which map to them. Suppose

$$\underline{v}_1 \longmapsto \underline{u}_1$$

$$\underline{v}_2 \longmapsto \underline{u}_2$$

$$\cdots$$

$$\underline{v}_p \longmapsto \underline{u}_p$$

If we can show that the set of vectors $B = \{\underline{k}_1, \underline{k}_2, \ldots, \underline{k}_m, \underline{v}_1, \ldots, \underline{v}_p\}$ form a basis for V, then the theorem is proved, for this will show that V has dimension $m + p$.

First of all, we show that any vector in V can be expressed in terms of these vectors, and then we show that B is a linearly independent set.

(i) Let $\underline{v}$ be any element of V; then we can find α_i's (real numbers) such that

$$L(\underline{v}) = \alpha_1\underline{u}_1 + \alpha_2\underline{u}_2 + \cdots + \alpha_p\underline{u}_p$$
$$= \alpha_1 L(\underline{v}_1) + \alpha_2 L(\underline{v}_2) + \cdots + \alpha_p L(\underline{v}_p)$$

i.e.

$$L(\underline{v}) = L(\alpha_1\underline{v}_1 + \alpha_2\underline{v}_2 + \cdots + \alpha_p\underline{v}_p)$$

i.e.

$$L(\underline{v}) - L(\alpha_1\underline{v}_1 + \alpha_2\underline{v}_2 + \cdots + \alpha_p\underline{v}_p) = \underline{u}_0$$

i.e.

$$L(\underline{v} - \alpha_1\underline{v}_1 - \alpha_2\underline{v}_2 - \alpha_3\underline{v}_3 - \cdots - \alpha_p\underline{v}_p) = \underline{u}_0$$

Thus $(\underline{v} - \alpha_1\underline{v}_1 - \alpha_2\underline{v}_2 - \cdots - \alpha_p\underline{v}_p)$ is in the kernel of L. There exist, therefore, numbers $\beta_1, \ldots, \beta_m$ such that

$$\underline{v} - \alpha_1\underline{v}_1 - \alpha_2\underline{v}_2 - \cdots - \alpha_p\underline{v}_p = \beta_1\underline{k}_1 + \beta_2\underline{k}_2 + \cdots + \beta_m\underline{k}_m$$

i.e.

$$\underline{v} = \beta_1\underline{k}_1 + \beta_2\underline{k}_2 + \cdots + \beta_m\underline{k}_m + \alpha_1\underline{v}_1 + \alpha_2\underline{v}_2 + \cdots + \alpha_p\underline{v}_p.$$

Thus any vector in V can be expressed as a linear combination of the vectors $\underline{k}_1, \underline{k}_2, \ldots, \underline{k}_m, \underline{v}_1, \ldots, \underline{v}_p$, i.e. B spans V.

(*continued on page 48*)

Solution 23.2.6.1 **Solution 23.2.6.1**

$$(A - B)^2 = A^2 - AB - BA + B^2$$

$$(A - B)^3 = A^3 - A^2B - ABA + AB^2 - BA^2 + BAB + B^2A - B^3$$

■

Solution 23.2.6.2 **Solution 23.2.6.2**

$$X = \begin{pmatrix} 0 & 0 \\ 0 & 1 \end{pmatrix} \text{ is an example.}$$

■

(*continued from page 47*)

(ii) We now show that B is a linearly independent set. Consider the expression

$$\beta_1\underline{k}_1 + \beta_2\underline{k}_2 + \cdots + \beta_m\underline{k}_m + \alpha_1\underline{v}_1 + \alpha_2\underline{v}_2 + \cdots + \alpha_p\underline{v}_p = \underline{v}_0 \quad (1)$$

Then

$$L(\beta_1\underline{k}_1 + \beta_2\underline{k}_2 + \cdots + \alpha_p\underline{v}_p) = \underline{u}_0$$

i.e.

$$\beta_1 L(\underline{k}_1) + \beta_2 L(\underline{k}_2) + \cdots + \alpha_p L(\underline{v}_p) = \underline{u}_0$$

i.e.

$$\alpha_1 L(\underline{v}_1) + \alpha_2 L(\underline{v}_2) + \cdots + \alpha_p L(\underline{v}_p) = \underline{u}_0,$$

because all the $\underline{k}$'s are in the kernel and so $L(\underline{k}_i) = \underline{u}_0$. But

$$L(\underline{v}_1) = \underline{u}_1$$

$$L(\underline{v}_2) = \underline{u}_2$$

$$L(\underline{v}_p) = \underline{u}_p$$

and the u's have been chosen to be linearly independent. Thus, all the α's are zero, and expression (1) reduces to

$$\beta_1\underline{k}_1 + \beta_2\underline{k}_2 + \cdots + \beta_m\underline{k}_m = \underline{u}_0$$

But the $\underline{k}$'s are also linearly independent, and so the β's are zero. Thus (1) implies that all the α's and β's are zero, and so the set is linearly independent, as required.

Unit No.		Title of Text
1		Functions
2		Errors and Accuracy
3		Operations and Morphisms
4		Finite Differences
5	NO TEXT	
6		Inequalities
7		Sequences and Limits I
8		Computing I
9		Integration I
10	NO TEXT	
11		Logic I — Boolean Algebra
12		Differentiation I
13		Integration II
14		Sequences and Limits II
15		Differentiation II
16		Probability and Statistics I
17		Logic II — Proof
18		Probability and Statistics II
19		Relations
20		Computing II
21		Probability and Statistics III
22		Linear Algebra I
23		Linear Algebra II
24		Differential Equations I
25	NO TEXT	
26		Linear Algebra III
27		Complex Numbers I
28		Linear Algebra IV
29		Complex Numbers II
30		Groups I
31		Differential Equations II
32	NO TEXT	
33		Groups II
34		Number Systems
35		Topology
36		Mathematical Structures